THE SOUP COOKBOOK

Enjoy Soups All Year Round with Easy, Delicious & Budget-Friendly Recipes Organized by Season and Ingredient | Includes Expert Tips & Secrets

TORI SACKLER

TABLE OF CONTENTS

INTRODUCTION

Delicious dishes have a timeless and adaptive creation. It has comforted many individuals across borders and fed the body and soul. The dish called soup is ordinary but outstanding. Soups, which are boiling pots of liquid seasoned with several ingredients, have been part of human cuisine for ages. Soups' delicious flavors, warm, calming environment, and health advantages make them crucial to a balanced diet.

This fascination with soups is a celebration of nutrition, wellbeing, and culinary ingenuity, not just flavor. There are many kinds and styles of soups, from chicken noodle and minestrone to Thai coconut and hot Mexican tortilla. Soups illustrate global food's complexity. All soups, regardless of origin, nourish, heal, and regenerate the body.

We'll search that sizzling dish for nutritional pearls to discover soup's health benefits. In addition to satisfying hunger, soups contain several essential nutrients, vitamins, and minerals. They also improve digestion, weight control, and hydration because they are liquid.

Conventional wisdom and modern scientific studies have always considered soups healthier. Soups are a diverse elixir for different nutritional demands and tastes. They can be used to treat a cold, fuel a busy day, or lose weight. These and more are soups' abilities.

We'll explore soups' various health advantages in later chapters. We will explore their immune system-boosting, weight-management, digestive system-calming, and nutrient-supplying properties. We will explore the world to find the regional versions of this beloved meal, each with its own history and healthy properties. Join us as we explore soups, a boiling stew of flavors and nutrition. Learn, inspire, and be amazed by the many reasons soups are essential to our culinary heritage and our quest for a better, more vibrant existence. Since ancient times, soups have helped us live healthier and more lively lives. Soups' health benefits will be revealed in the following pages like a slow-simmering broth that will warm your body and spirit.

CHAPTER 1

The Foundations of Soup-Making

The fundamental definition of the terms "comfort" and "nutrition" can be summed up in one word: soup. This culinary marvel that has endured across time and across civilizations. Soup holds a special place in both our hearts and our kitchens, whether it is consumed as an appetizer, as a meal in and of itself, or as a treatment for the common cold. But what are the fundamentals of preparing soup, as well as the tricks that go into making a bowl of liquid bliss that satisfies the soul? Let's investigate the fundamental components that make up the core of this age-old type of culinary art.

- The Fundamental Component: Broth or Stock

The liquid base, which can often be either broth or stock, serves as the dish's primary structural component. While bones are the primary ingredient in stock, meat, fowl, or fish can be used to make broth; however, stock is known for having a more robust flavor overall. Both contribute a sense of depth and character to soups, serving as a sort of canvas on which other ingredients can take center stage. It is of the utmost importance to choose a foundation that is of high quality, whether you are creating chicken noodle soup with a soothing chicken broth or a hearty beef stew with a robust beef stock.

- Vegetables and Herbs That Have an Aroma

In French cooking, onions, carrots, and celery are often used as the "holy trinity" to provide the flavorful base for a variety of soups. These vegetables, when sautéed in butter or oil, release their flavors and provide a pleasant sweetness to the broth. This is achieved by releasing the tastes of the vegetables. When it comes to seasoning and aromatics, herbs such as thyme, bay leaves, and parsley are absolutely necessary. Nevertheless, the selection of veggies and herbs might be very different depending on the kind of soup that you are making. For example, the cilantro and chili peppers that go into a tortilla soup are very different from the rosemary that goes into a substantial potato soup.

- The Highlight of the Recipe

Every bowl of soup worth remembering features a standout component that steals the show. It doesn't matter if it's the soft bits of meat in a beef stew, the smooth pureed pumpkin in a bisque, or the plump seafood in a chowder; the star ingredient is what determines the personality and flavor profile of the soup. Pick your shining jewel carefully, and then prepare it in a way that brings out its inherent beauty and appeal.

- Seasonings and other Spicy Ingredients

When preparing soup, flavoring it correctly is quite essential. The only required seasonings are salt and pepper, but the combinations are virtually limitless. Consider including spices like cumin, paprika, curry, or turmeric in your dish to create depth and complexity. The best spices to use will depend on your personal flavor preferences as well as the type of cuisine you are attempting to master. It is important to keep in mind that seasoning is a delicate balance; it is far simpler to add extra seasoning at a later time than it is to save an over salted soup.

- Texture and Consistency of the Ingredients

There is a wide range of textures and consistencies available in soups. Some are blended to the consistency of luxuriously smooth velvet, such as a traditional tomato bisque, while others are left lumpy and hearty, such as minestrone. Take into consideration the texture that you want your soup to have, and make any necessary adjustments using methods such as mashing, mashing with an immersion blender, or using a food processor.

- Complementary Elements and Final Touches

The skillful application of garnishes can transform an average bowl of soup into an unforgettable occasion. A sprinkling of grated cheese, a dollop of sour cream, a dollop of fresh herbs, or a drizzle of olive oil may all add visual appeal and bursts of flavor to a dish. Experiment with a variety of toppings to find one that works well with the flavor of the soup and the way it looks.

- Time and Patience

To make a good soup, time and patience are typically required. Simmering the ingredients over a low heat for an extended period of time helps the flavors develop and mingle. The soup becomes harmonious and well-balanced as a result of the slow cooking method that is used, which allows the essence of each component to be infused into the soup.

- Adaptability and originality come next in the list.

Even if traditional recipes offer a good starting point, don't be afraid to make adjustments or get creative with them. The preparation of soup is an art, and there are innumerable avenues open to be explored. Utilize foods that are in season, try out different regional cuisines, and trust your instincts when it comes to cooking.

The fundamentals of preparing soup are founded on a combination of fundamental cooking skills and a limitless capacity for creative expression. To achieve mastery in the art of making soup, one must have a solid understanding of these fundamental concepts, but they must also leave room for creativity and individual expression. The art of creating soup is a timeless and gratifying journey through the world of many flavors and textures. Whether you are making a bowl of chicken noodle soup for a sick loved one or building an exquisite bisque for a special occasion, the art of making soup is a journey that will never become obsolete.

Tools and Equipment

1. **A Sturdy Pot:** A good pot is the foundation of a successful soup-making experience. You will want one that is roomy enough to accommodate all of your ingredients without being cumbersome to work with, but not so massive that it overpowers your stove. Your best buddy in the kitchen is a heavy-bottomed soup pot because it prevents food from getting scorched and distributes heat evenly.

2. **Wooden Spoon:** For stirring and dishing, you should invest in a reliable wooden spoon or ladle made of wood. It won't scratch your pot, and while you're stirring a pot of goodies as it simmers, it has a nice, comfortable feel in your hand.

3. **Knife:** A Knife That Is Very Sharp Because making soups frequently requires a lot of chopping, you need a knife that is very sharp. The preparation work, such as chopping onions, slicing carrots, or mincing garlic, will go much more quickly

and easily if you have a nice knife.

4. **Cutting Board:** A robust and spotless cutting board is an absolute requirement for any kitchen. Choose one that has a cutting surface that is sufficiently large to allow you to comfortably chop your vegetables without the food sliding off the edge.

5. **Immersion Blender:** Although it is not required, having an immersion blender on hand when making creamy soups can be a huge advantage. You won't have to go to the hassle of pouring hot soup into a blender and running the risk of having it splash all over you.

6. **Strainer or Sieve:** In those instances, in which you desire a texture that is as smooth as silk, a strainer or sieve with a fine-mesh screen will assist you in accomplishing this goal by removing any lumps or sediments.

7. **Cups and Spoons for Measuring:** When it comes to cooking, precision is essential, and having cups and spoons for measuring will ensure that you receive the exact amount of each ingredient.

8. **Stockpot or Slow Cooker:** A stockpot or slow cooker comes in handy whether you want to make huge batches of food or just want to be able to set it and forget it. They are able to contain more soup and maintain the temperature for a longer period of time.

9. **Timer:** Because many soup recipes require the soup to be simmered for a predetermined amount of time, using a timer will allow you to keep track of the simmering process without having to continuously check the clock

10. **Tasting Spoons:** Your taste receptors are the most significant tool you will ever use. Make sure you have a few tasting spoons on hand so that you can taste your soup as it simmers and make any necessary adjustments to the spice.

11. **Containers for the Storage of Food:** Don't Forget About the Leftovers! Invest in some airtight containers so that you may keep your wonderful soup concoctions in the freezer or refrigerator without spoiling.

Now that you've got your arsenal of soup-making tools, you're ready to embark on a culinary adventure. From hearty stews to light broths, the possibilities are endless.

Types of Pots and Pans

Pots and pans, the workhorses of the kitchen that rarely get the recognition they deserve. When it comes to preparing a satisfying pot of soup, having the appropriate cookware can make all the difference in the world. Let's have a look at all the different kinds of pots and pans out there and figure out which ones are best for creating soup.

1. Stockpot: The Soup Maestro:

The stockpot is your best partner in the kitchen when you're trying to create hearty, family-sized pots of soup that will fill your home with delectable fragrances. Your mission: to warm your soul. These sturdy containers are the unsung heroes of the world of preparing soup. They are quietly standing ready to turn your culinary dreams into sizzling realities.

In the first place, stockpots are designed to be able to hold quantities that are sufficient to feed a little army. Because of their expansive depths and broad shoulders, they are easily able to make room for a vast quantity of liquid as well as a plethora of components. Whether you're making a rich beef stew to warm your bones on a cold winter night or a hearty chicken noodle soup to nurse a cold, the stockpot is there for you. It can do it all. However, size alone is not the only factor to consider. Because they are so good at maintaining a consistent temperature, stockpots are the best cooking vessels to use when you want to encourage the flavors in your dish to combine and become one. The trick to successfully distributing heat across the great majority of the liquid contained therein is to have a bottom that is both thick and heavy. There will be no hot spots, and nothing will be burned; simply a low, constant simmer that will take ordinary materials and turn them into something amazing. If you are on the market for a stockpot, you should keep an eye out for one that has handles that are solid and resistant to heat. These grips will give you the confidence you need to lift your bubbling mixture. As your soup slowly cooks to perfection, a lid that can be securely fastened and fits snugly over the pot is an invaluable ally. It helps to preserve the delicate tastes that have been carefully developed.

The stockpot is king when it comes to creating soup since it provides you with the perfect canvas on which to create culinary masterpieces that will bring comfort to people's hearts and satisfy their hunger. Therefore, whether you are creating a time-honored family recipe or beginning on a culinary adventure with exotic ingredients, reach for your stockpot and allow it to orchestrate a symphony of flavors that will leave your taste buds singing with joy. Whether you are forging a time-honored family recipe or embarking on a gastronomic adventure with unusual ingredients, grab for your stockpot.

2. Dutch Oven: The Versatile Workhorse

The Dutch oven, a workhorse of every kitchen, is like a seasoned traveler in the world of cooking. Like the fabled Swiss Army knife, these hardy, multipurpose pots can be used for a wide variety of purposes. What makes Dutch ovens the true unsung heroes of the kitchen? Let's peel back the layers and find out.

a. From Stovetop to Oven: A Seamless Transition: The Dutch oven is the undisputed king of multitasking, able to go from burner to oven with the flip of a switch. Envision first gently browning some aromatic onions and garlic in a pot, then transferring the whole shebang, soup pot and all, to the oven for a long, gentle simmer. The Dutch oven can do the work of several different pots at once.

b. Slow-Cooking Marvel: when it comes to slow cooking marvels like soups that beckon with rich, deep flavors. They are the undisputed kings of slow cooking thanks to their sturdy construction and thick walls. Dutch ovens have the patience and stamina to coax the absolute best from your ingredients, whether you're making a fragrant gumbo, a cozy chicken noodle soup, or a hearty minestrone.

c. Methods for Cooking Tough Cuts of Meat: Think of it: bits of tender beef, hog, or lamb giving victim to the Dutch oven's tenderizing powers over the course of a few hours. This culinary workhorse can handle tough portions like brisket, chuck, or shank with ease. They become tender, melt-in-your-mouth chunks that impart an unrivaled depth and richness to your soup when cooked over low heat with a tightly fitting lid.

d. The Art of Even Heat Retention: Dutch ovens are renowned for their ability to maintain an even temperature throughout the cooking process. The high walls and heavy lids of a pot do more than look good; they also serve to retain heat and disperse it uniformly inside. Put an end to burnt toast or soup that's too cold. The Dutch oven guarantees a uniformly pleasant and gentle simmer for all of the ingredients.

e. More Than Just Soup: We've been praising the Dutch oven for its soup-making abilities, but it has many other uses as well. You can use it to braise meat till it's fall-off-the-bone soft, simmer savory meals, roast a golden chicken, bake crusty loaves of artisan bread, and even deep fry with pinpoint accuracy. This appliance is a true multitasker, capable of meeting any cooking challenge you can throw at it. In a nutshell, the Dutch oven is an indispensable tool that goes above and beyond the call of duty in the kitchen. It's the vehicle by which ordinary items become gourmet treats. The Dutch oven is the unsung hero behind innumerable outstanding dinners, thanks to its ability to go from cooktop to oven with ease, slow cook like a pro, tame difficult portions of meat, and distribute heat evenly. Invest in one of these wonders of the kitchen if you want to take your cooking to the next level and experience a new world of flavors.

3. Saucepan: The Small-Batch Specialist

When it's just you and your significant other, the saucepan becomes your go-to pot for making comfort food from scratch. Miniature kitchen aids that combine efficiency and simplicity to perform culinary miracles on a more intimate scale were created with the accuracy of a soloist in mind.

Let's say you're in need of a bowl of hot, filling soup but don't feel like going to a lot of trouble making it. The saucepan now assumes center stage. You won't have to wait long for your soup fantasies to come true because to its small size and rapid heating capabilities. The cozy confines of the saucepan are the ideal setting for your culinary imagination, whether you're dreaming of a smooth and velvety bisque or a healthy and robust vegetable medley.

However, that's not all this handy saucepan can do. Repurposing old materials is equally as important as creating brand-new works of art. When the leftover soup from the day before calls to you from the fridge, you may bring it back to life with the help of a saucepan and some low heat so that it tastes as good as the day you made it.

The unique selling proposition of the saucepan is the degree of control it gives the cook. Its compact size allows for precise control, making it suitable for finishing touches like reducing a sauce or emulsifying a vinaigrette. The saucepan is the conductor of little culinary symphonies in the hands of a skilled cook.

Whether your culinary soul yearns for the sophistication of smooth bisques or the ease of pureed vegetable soups, the reliable saucepan will become your go-to tool. Let the saucepan do its culinary magic, generating a symphony of flavors that will make every meal a memorable encore, whether you're cooking for yourself, a treasured companion, or just seeking to relish a small-batch delight.

4. Sauté Pan: The Soup Starters

When it comes to making a truly mouthwatering soup, the sauté pan serves as the equivalent of an opening act, preparing the groundwork for the spectacular gastronomic show that is about to take place. Even if it might not be the star of your soup-making show, its part in creating the flavor is quite essential. Even if it isn't the star, its job is absolutely necessary.

Because of their low, sloping edges and vast cooking surface, sauté pans are the unsung heroes of the kitchen. They are the ones who are responsible for coaxing out the depth and diversity of flavors in your soup. Their expertise is in searing meats and veggies to a golden brown to the point of perfection. Imagine the following: the meat is scorching to a golden-brown crust, which locks in the meat's fluids and amplifies the flavor; the onions sizzle as they reach the hot pan, releasing their sweet and savory aroma; the sizzling of the onions as they touch the hot pan.

The enchantment, however, does not end there. Aromatics such as garlic, ginger, and spices are brought to life in sauté pans, where they expel their essential oils and permeate the air with their enticing aromas. It is the first movement of the symphony, the prelude to the simmer, and it is at this time that the contents convert from being simple components into the fundamental elements of a flavorful soup. Despite the fact that sauté pans may not have the depth necessary to accommodate all of the liquid and components required for your soup, their job is nonetheless essential. After the browning and sautéing stages have been completed, it is time to make the transfer. This involves making a deft handoff to the stockpot or Dutch oven, where the sautéed components will join forces with broths, stocks, and other seasonings to begin on the trip of a prolonged, slow simmer.

Therefore, keep in mind the importance of the sauté pan the next time you go on the journey of creating soup. It is the narrator of the flavor tale of your soup and the key to those deep, complex flavors that will leave your visitors guessing about your culinary skills. It is the point at which the flavoring process begins, and it is here that you will make certain that each and every bite of your soup is a symphony of flavors and sensations.

5. Soup Pot: The Soup-Specific Star

Soup lovers, rejoice! The soup pot was made for you, and it's all you need to achieve culinary nirvana. These carefully built containers are like the Stradivarius of the soup world; they are precisely tuned and designed for one thing only: to give life to your soup creations.

Soup pots should be able to serve multiple purposes. These pots come in various sizes to satisfy your soup-related whims and fancies. There's a soup pot out there that'll be perfect for your next dinner party, whether it's just you and your significant other or a large group of friends.

The quality of design put into these jars sets them apart. They have roomy interiors so that a light stir may distribute ingredients evenly and bring out a dish's full taste. They have mastered the art of simmering, the low, continuous heat necessary for the magic of broth, herbs, and ingredients to come together in perfect harmony.

Some soup bowls go to extremes for the soup they make. Envision a pot with an integrated strainer, ready to skim the flavorful broth from the hearty solids with a simple tilt of the handle. It's like having a sous chef who can anticipate your needs and jump in at the right time to simplify your cooking.

Cooking a pot of soup is like conducting a symphony of flavors, scents, and textures. These vessels are your conductor's baton as you lead your components to a delicious climax in the kitchen. The next pot of soup you make will be a masterpiece, as they respect tradition while promoting your innovative cooking.

Soup pots are essential tools for any cook, whether they're preparing a comforting batch of chicken and rice or experimenting with an exotic blend of ingredients. It's more than just a vessel for holding liquid; it's an invitation to the delicious and historic realm of soup.

If you are the type of foodie who views the preparation of soup as something that straddles the line between an art and a science, then copper pots should have a prominent place in your kitchen armory. These glistening pots are not just for show; rather, they are the ideal of precision and responsiveness, and they are designed to take your soup making skills to an entirely new level.

The unrivaled ability of copper pots to conduct heat is one of the material's defining characteristics. Copper, in contrast to many other materials, demonstrates a nearly immediate response to variations in temperature. You are able to fine-tune your cooking process with extraordinary precision thanks to the copper interior of your pot, which responds with remarkable agility whenever the heat beneath your pot is adjusted.

Because copper pots are so responsive to heat, they are ideal for making delicate soups that call for a nuanced touch, such as velvety seafood bisques or silky-smooth cream-based compositions. Copper gives you the ability to exert influence over every stage of the preparation of your soup, from the tender sautéing of aromatic shallots to the exacting reduction of a broth that has been infused with wine.

In spite of this, there is a catch that one has to be aware of before embracing these precision tools of gastronomic delight. To keep their gleaming appearance, copper saucepans require some care and attention. They eventually acquire a natural patina that, over time, can have an impact on how well they function. Polishing them on a regular basis is required in order to maintain their stunning appearance while they are in use.

Therefore, if you are ready to put in a little more effort to keep your copper pots in great condition, they will reward you with a level of control and elegance that may transform your soup-making activities into a work of culinary art. If you are not willing to put in the extra effort, your copper pots will not reward you with the degree of control and refinement that you deserve. When you have copper working for you in the kitchen, each and every bite of soup will be a representation of your unwavering commitment to achieving culinary perfection.

6. Copper Pots: The Precision Picks

Copper pots are the peak of precision and artistry for the soup-making connoisseur who views their craft as an art form. These shiny pots and pans are more than just cooking tools; they're musical instruments that can help you create a culinary symphony.

Those who enjoy their soup very much tend to favor copper pans. Their ability to conduct heat is their trump card. They're like the race cars of the kitchen, reacting to changes in temperature with incredible speed and accuracy. Copper pots are your best friends when it comes to making silky seafood bisques or delicious consommés. They make sure the soup is simmered just right so that all the flavors can develop and the aromas may waft out.

But copper pots require some care and attention, just like any culinary genius. They need to be polished now and again to keep that lustrous, shining brilliance that makes your kitchen the envy of the neighborhood. With careful care, they can become heirlooms and be passed down from one generation of cooks to the next, creating a patina that symbolizes their culinary adventure along the way.

Copper pots are the precision selections that bring your dishes to a level of elegance and taste that is nothing short of outstanding, so they are the obvious choice for people who seek culinary excellence and who relish each mouthful of soup as if it were a work of art. They will make you feel like you're orchestrating a symphony of tastes in the kitchen, and your visitors will be in awe of your culinary skills.

7. Nonstick Pots: The Easy-Clean Option

The stockpot becomes a reliable partner in your kitchen when you want to make a lot of hearty soup at once. These huge, deep pots can easily accommodate a lot of liquid and ingredients. The stockpot is the workhorse of the kitchen, indispensable whether you're making a luscious broth, a robust stew, or a fiery chili.

The Benefits of Using Nonstick Cookware

Let's talk about how nonstick pans are the unsung heroes of easy food preparation and cleanup. When it comes to making tasty soups with minimal cleanup, nonstick pots are your best bet if you prioritize efficiency in the kitchen.

Ceramic or PTFE (polytetrafluoroethylene) coatings on the inside of nonstick pans are what make them so effective. These coatings produce a slick surface that prevents food from adhering to the pot's walls, facilitating easy washing. After all, it would be a waste of time to brush off dried soup when you could be eating. But there are a few things to remember when using nonstick pots, as with other culinary item, to make sure they last for years:

Use Only Soft Utensils: Wood, silicone, or plastic utensils are safest for use with nonstick cookware. Using metal utensils can damage the nonstick coating and reduce its useful life. You shouldn't use any metal utensils, including spoons and whisks, in your nonstick pot.

Nonstick pots are great since they distribute heat uniformly, but they shouldn't be heated to high degrees. Turning up the heat can destroy the nonstick coating, so it's best not to do that. When preparing soup, use a stovetop burner set to medium or medium-low heat.

When cleaning, use mild detergents and soft sponges or dishcloths to avoid scratching surfaces. The nonstick coating can be worn down over time by using abrasive scouring pads or aggressive cleaning solutions.

Soups like thick chowders and creamy bisques benefit greatly from being cooked in nonstick pots because they don't adhere to the surface. The soup won't stick to the spoon or the pan thanks to their easy-release qualities.

When venturing out on culinary adventures, keep in mind that your nonstick pot is a reliable companion that will cut down on your prep time and cleanup so you can focus on enjoying your soups. If you're kind to it, it'll keep on giving you hot soup until you're sick of it.

Essential Ingredients

1. Broth or Stock:

Let's get down to the good stuff—the broth or stock that forms the basis of so many delicious soups. This precious metal is the unsung hero, the solid base, and the indispensable cornerstone upon which the unique flavors of numerous soups are built. A well-made, high-quality broth or stock is like having the key to a treasure trove of culinary possibilities, whether your preference lies with the warm embrace of chicken, the robust depth of beef, the earthy warmth of vegetables, or the briny essence of seafood.

Imagine an alchemical process starting with a pot of bubbling potion and some aromatic ingredients. As the veggies break down and the proteins infuse, the broth or stock you've chosen becomes the symphony's maestro. Its lusciousness is like a warm embrace, its savoriness like a dance on your taste receptors, and its comfort like a gourmet cuddle.

It's not merely a liquid but rather the liquid itself. The magic potion that can make even the most basic of materials into a nourishing soup, stew, or broth. It's the basic flavor that brings back memories of home cooking and soothes the minds and bodies of people who indulge.

So, when you go out on your soup-making adventures, keep in mind that the broth or stock you select will serve as the base upon which the rest of your soup will be created. Pick carefully; it's the foundation, the soul, of your soups' journey to the stratospheric heights of flavorful deliciousness.

2. Fresh Vegetables:

There are certain soup recipes that will never go out of style, and those are the classics. The carrots in your broth add a soothing sweetness thanks to their sweet earthiness. The celery adds a light, herbal flavor, while the onions and garlic work together to create a robust scent that will permeate every bite of your soup. Tomatoes should not be forgotten either; their acidity and richness add a wonderful dimension to many a favorite dish.

However, soup's adaptability is part of its charm. The kitchen is your laboratory; don't be afraid to try new things and add your own unique flair. Picture some bright red bell peppers adding sweetness and a hint of smoke to your stew. Leeks, which have a mild onion-like flavor, tango between the realms of onions and garlic, giving your dish a certain allure all its own. And if you're looking to give your soup a velvety texture and nutty, creamy undertone, butternut squash's warm embrace is a wonderful addition.

When it comes to making soup, the only limit is your creativity. So, whether you're making a classic or trying something new in the kitchen, remember that fresh vegetables can take your soup to a whole new level of flavor.

3. Herbs and Spices:

Be careful to keep a wide variety of these enticing items on hand in your well-stocked pantry. Sweet and fragrant basil can add a hint of Mediterranean heat to your soups. Oregano's bold, Mediterranean flavor can transform even the simplest broth into a scrumptious meal, while the earthiness of thyme adds a calming, woodsy aroma to your preparations.

Then there are bay leaves, which stand guard as one of the most powerful herbs around. Throw one into your pot of boiling food and watch as it creates a strange and rich tapestry of flavors that will leave your senses captivated.

But we mustn't forget salt and pepper, the real heroes of seasoning. Because of salt's flavor-boosting properties, every element of your soup will blend together in delicious harmony. Pepper, on the other hand, can wake up even the doziest of taste senses with its mild heat and hint of earthiness.

So, when you step forth into the world of soup-making, remember to always utilize your herb and spice collection to its full potential. They are the sorcerers that work their magic in your kitchen, transforming ordinary ingredients into delicious treats.

4. Pasta and Grains:

Pasta is a fantastic component since it can add a variety of textures to your soup, and it comes in a wide variety of shapes and sizes. Pasta, whether it's angel hair, penne, or bowties, takes up the taste of the broth and provides a pleasant chewiness to each bite. It adds weight to the dish, but it also brings a sense of familiarity and warmth that makes you think of mom's cooking.

The grain of rice in your soup will act like little flavor sponges, soaking up the broth while adding a soft, delicate crunch. Rice, whether it's long or short grain or an aromatic variety like jasmine or basmati, adds a creamy texture to your bowl that's like being enfolded in a warm blanket on a cold day.

Barley is a classic ingredient for enhancing the richness and depth of soups with its nutty, meaty flavor. It's well-known for releasing a delicate, earthy sweetness while it cooks, giving your soup a hearty, rustic flavor. Barley improves the soup by adding a chewy, robust quality.

Toss in some quinoa for a healthy dose of protein and a fun new texture in your next pot of soup. When cooked, its tiny, bead-like grains remain pleasantly hard, making a nice contrast to the soft veggies and savory broth. For a healthy and hearty soup, quinoa is an excellent option due to its slight nuttiness and intrinsic vitamins.

Using these pasta and grain substitutions in soup recipes is like creating a beautiful tapestry of tastes and textures. They add to the satisfying texture of your soups and help create a homey, comfortable meal that's good for the body and the soul. Pasta and grains are your tried and true companions on this culinary adventure, whether you're cooking up a pot of chicken noodle soup, a bowl of minestrone, or a quinoa and vegetable medley.

5. Proteins:

Proteins are soups' souls and open the way to global cuisine. A selection of protein options in your pantry or fridge is like a passport to a number of soups, whether you're a carnivore, seafood lover, or plant-based eater.

> **Chicken:** Classic chicken adds homey taste to soups. Chicken goes nicely with many components, from tender shreds in a robust chicken noodle soup to juicy bits in a creamy chicken and wild rice dish.

> **Beef:** For meaty, rich flavors, beef shines. Consider the rich, savory flavor of beef stew or the comforting warmth of beef and vegetable soup. It's the best soup for depth and richness.

> **Seafood:** Seafood lovers have many options. Enjoy a creamy shrimp bisque or clam chowder with delicate clams and crunchy bacon. Seafood adds marine flavor to soup.

Plant-based? Try tofu. Tofu is your protein choice. Silken tofu makes a velvety basis for vegan tomato bisque, while extra-firm tofu marinates and pan-fries to provide structure and protein to Asian-inspired soups.

These different protein options let you discover soup's rich flavors, textures, and cuisines. So load up your pantry and fridge and let your taste buds explore your fantasy soup cookbook.

6. Dairy or Dairy Alternatives:

The decision between dairy and its replacements becomes a critical point in your culinary adventure when it comes to creating soups that are not only delicious but also accommodate to diverse dietary needs. You can achieve gastronomic nirvana with the addition of their creamy flavor to your soups.

Cream and milk, two staples of the dairy world, give soups a luxurious, silky texture. The addition of cream's sumptuous richness can elevate a simple soup to the level of a velvety masterpiece. In contrast, milk provides a softer touch of decadence with a smoothness that is no less enticing for being more understated.

There is a plethora of alternatives to dairy for those who are ready to take the plunge into dairy-free soup magic. The subtle nuttiness of almond milk makes it a suitable substitute for regular milk, and it also adds a welcome new dimension to your culinary creations. But the silky, subtly sweet flavor of coconut milk is just what you need to give unique recipes a taste of the tropics. Cashew cream is a delicious vegan option that can be used to add a rich, buttery touch to your baked goods.

Your soups will thank you for the addition of creamy, dreamy sweetness, whether you follow the traditional dairy route or pick one of these tantalizing alternatives. Get creative in the kitchen and see what works best for you!

7. Aromatics:

Aromatics are those lovely components that fill your soups with enticing aromas and complex flavors. Aromatics are also known as flavor enhancers. Ginger and lemongrass are two aromatic ingredients that really stand out when it comes to Asian-inspired soups. A symphony of flavors is created when the zesty, citrusy notes of lemongrass and the warm, earthy spiciness of ginger combine in perfect harmony to produce a dish. This dish will take your taste buds on a journey to the bustling streets of Southeast Asia. These aromatic powerhouses are your passport to a world of culinary delight, whether you're cooking up a soothing bowl of Vietnamese Pho or a heady bowl of Thai Tom Yum.

And when it comes to conjuring the exquisite tastes of Middle Eastern cuisine, the fragrant combo of cumin and coriander takes center stage as the star players. A dish like Moroccan Harira soup or Lebanese lentil soup is given an additional layer of complexity by the addition of cumin, which has a flavor that is warm and slightly nutty. Coriander seeds, however, add a subtle and lemony allure that lends a delectable twist to your Middle Eastern recipes, making them really unique and unforgettable. Each little scoop of these fragrant ingredients is like taking a trip to the lively souks of Marrakech or the throngs of people on the streets of Istanbul. Therefore, you should not be afraid to embrace the magic of aromatics and allow them to take you on a gastronomic adventure that transcends multiple continents and civilizations.

8. Fresh Herbs:

Imagine a steaming cup of soup that tantalizes your senses with its enticing aroma and fresh herbs. The soup is scorching hot and the steam is rising. Now, consider the enchantment that takes place just before you take that first bite, when you delicately sprinkle a handful of fresh herbs that are vivid and fragrant over the surface of the dish. It's like a torrent of natural goodness gushing into the beauty that you've created in the kitchen. This herbaceous symphony is finished off with the mellow, oniony fragrance of chives, which complements the earthy warmth of parsley like a perfect dancing partner. The emerald leaves of cilantro, with their bright, citrusy undertones, are the other key player. These fresh herbs are not just a garnish; they are the final brushstrokes on your soup canvas, adding a burst of color, scent, and flavor that transforms each spoonful into a gustatory masterpiece, a celebration of nature's wealth in each bite. In other words, they are not just a garnish; they are the final brushstrokes on your soup canvas.

9. Stock Up on Canned Ingredients:

Gather a Supply of Canned Goods for Your Kitchen: It's undeniable that our lives may become quite chaotic at times, which is precisely why the convenience aspect is so important. When time is of the essence but the need for a warm bowl of soup is through the roof, having a well-stocked pantry with a variety of canned necessities can be a lifesaver.

A dependable can of tomatoes is like a kitchen superhero. Countless soups, from tomato bisque to robust chili, start with them. They give your food an unrivaled depth and intensity of taste. It's like eating a little ray of sunshine.

Protein-rich and convenient, canned beans are a cupboard staple. Canned beans offer heartiness and nutrition to any soup, whether you're making a traditional minestrone, a fiery chili con carne, or a speedy Mexican-inspired tortilla soup. They're like tasty little morsels just ready to make your soups more filling.

Corn from a can is sweet, crunchy, and always at the ready, making it a fantastic addition to any soup. Chowders, vegetable soups, and southwestern-inspired dishes benefit greatly from its use because its natural sweetness balances the savory tones of your broths. Even on the coldest days, it's like a ray of sunshine.

Whether you're in a need to get supper on the table or just want to comfort yourself with some homemade soup, these canned goods will come in handy. They are the magic ingredient in the kitchen that may make your stomach full and your nerves quiet. When you keep canned goods like tomatoes, beans, and corn on hand, you can whip up a pot of hot, hearty soup in no time.

10. Extras for Garnish:

Garnishes allow you to express your individuality as a chef when it comes to the last touches of your soup. These tasty additions not only make your bowl of comfort food seem more appetizing, but they also enhance the taste and texture of your soup. Explore the various outcomes with me.

The addition of grated cheese to soup is like a soft snowfall. A sprinkling of cheese, whether it's old cheddar for a sharp tang, Parmesan for creamy luxury, or Gouda for smokey charm, melts into your soup and imparts a rich and savory aroma that's hard to resist. Croutons are the perfect complement to a bowl of soup because of the delightful crunch they add to the otherwise silky broth. To add a homey touch to your dish, you may either create your own croutons from day-old bread and season them with herbs and spices, or you can buy a bag of your preferred artisanal croutons.

A little bit of good olive oil drizzled on top of your soup may make it taste like something straight out of the Mediterranean. It's like a ribbon of liquid gold that floats through your soup, giving it a velvety texture and a light fruity, peppery flavor. Because of its rich flavor, extra virgin olive oil shines brightest in this position. Natural confetti for your soup, fresh herbs are a wonderful addition. They are a delicate addition to any dish, adding a pop of color and a new dimension of flavor. A dash of any of these green miracles, from basil for a touch of sweet aromatic notes to parsley for an earthy freshness or chives for a mild oniony zing, gives your soup a garden-fresh dimension.

Add some zip to your soup by grating the zest of a citrus fruit, like a lemon or lime, over top. The citrus oils that are released provide a refreshing, acidic kick to your dish by waking up your taste buds and brightening the overall flavor profile.

Toss in some toasted nuts or seeds for an earthy, nutty crunch and a change in texture. A range of soups, from light cream soups to robust stews, benefit from the addition of almonds, pine nuts, or pumpkin seeds for their delicious crunch and subtle nutty taste. Creamy, tart sour cream or yogurt is a great addition to your soup, especially when stirred in at the last minute. It reduces the spiciness of soups and gives cream-based soups a luxurious, velvety texture Crispy bacon bits are the perfect finishing touch, and they add a delicious crunch. Their smokey, salty charm not only adds an appealing layer of taste, but also produces a fascinating difference in texture, turning every bite into a savory delight.

Don't forget that there are no rules when it comes to the art of soup garnishing beyond your own creativity and personal choice. You can take your soup from ordinary to extraordinary by adding a variety of garnishes.

Preparing Vegetables

Vegetable prep is essential for making a healthy and filling soup. How you treat and cook your vegetables is crucial, whether you're creating a substantial minestrone, a creamy potato leek soup, or a light and refreshing gazpacho. In order to become an expert vegetable soup preparer, consider the following advice.

Washing and Cleaning:

1. Make sure your sink is spotless and clear of any debris or dirt. Remove any lingering residue or grime by giving it a brief rinse in hot water.

2. In making vegetable soup is to gather all the ingredients and set them out. Before moving forward, make sure there are no obvious evidence of dirt, insects, or damage.

3. Run Cold Water Over Your Vegetables While They're Laid Out Under the Faucet. Vegetables will keep longer in cold water, extending their shelf life. Gently rub and rotate each veggie under flowing water to provide a thorough cleaning.

4. For veggies like carrots and potatoes that have thicker skins or surfaces that may harbor dirt, a vegetable scrub brush is an effective tool for removing grime. With one hand, firmly grasp the vegetable, while the other holding the brush, gently scrub the surface under running water. To prevent dirt from collecting, give uneven surfaces and cracks your full attention.

5. When handling tomatoes, peppers, or leafy greens, it's better to use your hands because of their delicacy. Take a cupful of water and pour it over the leaves or veggie. To remove dirt from the surface, give it a light rubbing. Take care not to squish or otherwise spoil the food.

6. Greens Extra caution must be exercised while working with leafy greens such as spinach and kale. Remove the leaves from the stems, as the latter may harbor more grime than the former. Put the leaves in a big bowl of cold water and add them one by one. The dirt can be washed away by swishing them in the water. The water's clarity may temporarily decrease.

7. Rinse Under Clear Water: Keep Doing This Until the Water Becomes Clear. Especially with leafy vegetables, this may require multiple rinses and swishes. The cleanliness of your vegetables can be judged by the transparency of the water.

8. After washing your vegetables, drain any extra water by gently shaking them. If you're worried about the texture of your soup being compromised by too much moisture, pat the vegetables dry with a clean kitchen towel or some paper towels.

By removing any grit or contaminants, the effort spent washing and cleaning your vegetables not only makes your soup safer to eat, but also improves its flavor. Now that the vegetables are ready, you may continue working on your delicious soup. Have fun experimenting in the kitchen!

Peeling:

Vegetables like carrots, potatoes, and butternut squash require peeling before they can be used in a soup. The dish's flavor and texture will change drastically depending on whether you peel the ingredients or not. In this article, we'll discuss the ins and outs of peeling various vegetables and why your method of choice matters.

- **Carrots:** Carrots, with their sweet and earthy flavor, are a common ingredient in many soups. It's up to personal preference and the kind of soup you're creating as to whether or not you should peel them. A few positives and negatives are as follows

Peeled Carrots: Using a knife or a vegetable peeler, you may give your soup a polished, professional look. The skin's natural bitterness is also neutralized in this process. Soups like bisques or those with cream are perfect candidates for using carrots that have been peeled. Some people prefer their carrot soup with the skins left on because it gives it a more earthy flavor. The skin adds a nutritious boost to your cuisine because it is packed with beneficial minerals and fiber. Soups with a rustic flavor, like minestrone or a hearty vegetable stew, benefit from having the skin left on.

- **Potatoes:** Potatoes, with their hearty, creamy texture, are another popular addition to soup. Soup flavor might vary depending on whether the vegetables are peeled or not.

Soups with a creamy potato or vichyssoise texture benefit from potatoes that have been peeled. This process also eliminates any potential skin-related off-flavors. Peeling is the way to go if you want to improve the feel and look of your skin. Keep the potato skin on for a more rustic flavor in your soup by using unpeeled potatoes. The skin enhances the dish with a lovely earthiness and a somewhat chewy texture. In addition, vitamins, minerals, and fiber are preserved. Unpeeled potatoes are great for making soups with a more rustic, homey feel.

Because of its natural sweetness and brilliant color, butternut squash is frequently used as an ingredient in various soup recipes. This is one reason why it is so popular.

- **Butternut Squash, Peeled:** Butternut squash is often best when peeled. Tough and harsh skin might detract from the flavor of your soup. To get to the sweet, soft flesh underneath, peel it with a vegetable peeler or a knife for a tastier and more appetizing end product.

The decision to peel veggies for soup comes down to personal preference and the final product you're hoping to achieve. Some soups benefit from having a smoother texture and a cleaner appearance, which can be achieved by peeling the ingredients. On the other hand, leaving the skin on can offer earthy flavors, rustic textures, and increased nourishment, which can enrich the character of heartier soups. The decision is yours to make, and either method can be experimented with to create new and interesting soups.

Chopping and Dicing:

Vegetable preparation may appear simple at first, but the quality of your soup will depend greatly on how well you chop and dice the vegetables. The speed and consistency of cooking are both affected by the size and uniformity of your cuts, both of which have an impact on the soup's texture. Let's get into the nitty-gritty of preparing soup ingredients.

1. **The Size and Texture of Vegetables:**

Chopping veggies into large pieces is the best option if you like your soup to have a chunky texture and big bits of vegetables. Cut your vegetables into larger chunks, often between 1 and 2 inches in size, to do this. Because of the greater size of the pieces, the veggies in your soup will retain some of their natural texture. Minestrone, chili, and chunky vegetable soups all benefit from having their vegetables chopped up.

However, if you want a more uniform texture and the tastes to blend together more seamlessly, chopping your vegetables into small, uniform pieces is the way to go. To dice, cut your vegetables into uniform cubes of 1/4 inch to 1/2 inch in size. With everything the same size, your soup will simmer evenly and have a smoother, more polished texture. Vegetables that have been diced perform particularly well in pureed soups, creamy bisques, and light broths.

2. **Tools and Methods:**

One of the most important tools in the kitchen is a sharp knife, which can be used for both chopping and dicing. Crushing and damaging the vegetables with a dull knife might result in uneven slices and a less appetizing texture. Purchase a high-quality chef's knife and maintain it with consistent sharpening.

The stability of the cutting board is also crucial. You should work with a stable cutting board that won't move around. Most people use either wooden or plastic cutting boards, but it's important to keep them clean and sanitized to avoid spreading germs.

3. **Think About the Type of Soup:**

Chopping works wonderfully in heartier soups like stews and chiles. The longer cooking times help the larger vegetable pieces retain their shape and flavor.

Dicing is recommended while preparing creamy soups like potato leek or butternut squash. Vegetables with uniformly small pieces cook more quickly and combine more easily, creating a velvety texture.

4. Cooking Time:

Chopped veggies, being larger in size, may need additional cooking time to soften and become tender. If you like a little "bite" and texture in your soups, this is the way to go.

Cooking time and temperature are both reduced when veggies are diced. This works wonderfully in soups when a smooth, even consistency is desired.

The key to making great soup is perfecting the art of cutting and slicing veggies. Which method you employ depends on the soup you're making and the consistency you want. With a sharp knife, a stable cutting board, and an understanding of your desired soup texture, you'll be well on your way to creating soups that delight the senses and warm the soul. Happy chopping and dicing!

Minced and Crushed Garlic:

Garlic is a flavoring powerhouse that is used in a wide variety of soup recipes; let's delve deeper into the art of working with garlic.

To easily add the powerful flavor of garlic to your soups, minced garlic is a great ingredient to have on hand. Follow these steps:

The Garlic Selection Process: Select fresh garlic bulbs as a starting point. Try to find bulbs that are solid and have skin that is not cracked. Soft or growing garlic should be avoided.

To get to the cloves, peel the garlic: Remove the cloves from the bulb. To remove the skin from a clove, put it flat side down on the blade of a chef's knife and press down gently but firmly until you hear a little snap. After that, peel it off. A silicone garlic peeler is another option that can make your life much simpler.

Garlic Mincing: Chop the garlic into little pieces. Prepare the clove by slicing thinly across its width. Take care not to cut yourself; avoid reaching for the blade with your fingers. Turn the garlic 90 degrees after slicing it and cut it very finely. The soup's garlic flavor will be more uniform if the garlic is chopped finely.

If you want a more subtle and well-integrated garlic flavor in your soup, try adding minced garlic early on in the cooking process. To lay a foundation for flavor, minced garlic can be sautéed in a little oil or butter at the start of your soup-making expedition.

Garlic, crushed:

Crushing garlic is another great method for improving the taste of soup. Here's the procedure:

To choose and peel new garlic cloves, simply repeat the procedures outlined in the previous section.

Crush the garlic with a garlic press, a useful kitchen appliance that does just that. Insert a peeled clove into the press's chamber, then firmly squeeze the tool's handles. Small pores in the device let the crushed garlic out, but the pulp stays inside.

Crushing garlic results in a different metamorphosis than mincing it does in terms of flavor release. Crushing garlic causes the cells to burst, releasing more of the plant's flavorful oils and fluids. Your soup's garlic flavor will be more intense and noticeable as a result.

Including Crushed Garlic: Crushed garlic is commonly added to soups when a strong garlic flavor is desired. Soups like garlic and potato, or a traditional tomato bisque, benefit greatly from its use. Crushed garlic should be added at the end of cooking so that its intense flavor is not lost.

Soups benefit greatly from the addition of fresh garlic, which can be prepared in a variety of ways (minced, crushed, etc.). Try both methods out and see which one yields the greatest results for your soups based on your own preferences and the desired flavor profile. Don't be afraid to play around with this aromatic ingredient; garlic is a culinary treasure that can take your soups to the next level.

Onions:

In many soups, onions play a central role because of the depth and complexity of flavor they add to the broth. Knowing how to correctly prepare onions is crucial if you want to get the most out of them when creating soup. If you're making soup, here's a comprehensive guide to dealing with onions:

Choosing the proper Onions: Choosing the proper onions is the first step in making a successful soup. Onions of all colors—yellow, red, and white—are available and often used. In most soups, yellow onions are the best option because of their mild flavor and versatility. White onions have a gentler, more delicate flavor than red ones, while red onions add a subtle, somewhat sweet aroma. Picking onions without thinking about the soup as a whole is a recipe for disaster.

Preparing Your Equipment:

Gather a sharp chef's knife, a sturdy cutting board, and a kitchen towel for cutting onions quickly and easily. Crushing the onion cells can release even more unpleasant substances, which can make you cry when chopping.

Getting the Onion Ready:

Start by chopping off the onion's stem and root ends. This makes it easier to cut through the papery skin and get to the meat underneath.

After slicing in half, place the onion on one of the flat ends. Cut the onion in half lengthwise, following the onion's natural curve. The layers will stay together better if you cut through the root end last.

After cutting the onion in half lengthwise, pull the skin off carefully. If you want to make vegetable broth, preserve the skin or throw it away.

It is up to you to decide whether you want to slice or chop the onion for your soup. Thin, semicircular slices of onion create a meaty, rustic feel. In a pureed soup, for example, the onion should be chopped very finely to get a smooth texture.

When it comes to enhancing the flavor of onions in soup, caramelizing them is one of the most delicious methods. Slowly frying thinly sliced onions over low heat causes them to develop a rich golden-brown color. Here's what you need to do:

To get started, heat a heavy pan over medium heat with a bit of oil or butter. Caramelizing onions is best done in olive oil or butter.

After the oil or butter has heated but before it begins to smoke, add the cut onions. Coat them evenly with the fat by stirring.

Lower the Heat: Reduce the heat to low or medium-low, ensuring the onions cook slowly and evenly. Stir occasionally to prevent sticking or burning.

Be Patient: Caramelizing onions is a slow process that can take anywhere from 20 to

45 minutes, depending on the quantity of onions and your desired level of caramelization. Stirring every few minutes will help achieve even browning.

Deglaze (Optional): If you want to intensify the flavor, you can deglaze the pan by adding a splash of broth, wine, or water and scraping up the browned bits from the bottom of the pan. This liquid can be added directly to your soup for an extra layer of flavor.

Season: Towards the end of caramelization, season the onions with a pinch of salt and a dash of pepper to taste. This step enhances their sweetness and depth of flavor.

Caramelized onions lend a rich, sweet, and savory note to your soup, making it more complex and satisfying. Whether you're preparing a classic French onion soup or looking to elevate the flavor of your favorite broth-based soup, mastering the art of caramelizing onions is a culinary skill worth acquiring. So, the next time you're crafting a soup recipe that calls for onions as the base, follow these steps to prepare and potentially caramelize your onions, and watch as they transform your soup into a culinary masterpiece.

Tomatoes:

If your soup calls for fresh tomatoes, blanch them in boiling water for about 30 seconds, then transfer them to an ice bath. The skins will easily peel off. Remove the seeds and chop the flesh to use in your soup.

Selecting Fresh Tomatoes: A tasty soup starts with selecting the correct tomatoes. Choose ripe, firm tomatoes with smooth, undamaged skins. When choosing tomatoes, consider your recipe and personal taste preferences. Different types have different sweetness and acidity.

Tomato blanching: Bring a kettle of water to a boil. You need enough water to submerge tomatoes. The quantity of tomatoes you use may require batching.

Scoring the Skin: Cut a shallow "X" or cross-shaped cut on the bottom of each tomato with a sharp knife as the water heats up. This will split the skin during blanching and make peeling simpler.

Blanching Time: Carefully lay tomatoes in boiling water using a slotted spoon or tongs. Blanch for 30 seconds. Skin will pull away from your cut.

Ice Bath: Place tomatoes in ice water soon after blanching. The rapid chilling procedure, called "shocking," stops cooking and retains tomato texture and flavor. Let them soak in the ice bath for two minutes.

Peeling Tomatoes: 5. Peel with Ease: After soaking in ice, the skins will relax and peel away from the flesh. Peel the skin gently with your fingers or a knife. Easy removal is expected.

Remove Seeds After peeling the tomatoes, remove the seeds. Horizontally cut tomatoes in half. Carefully strain the seeds and juice from each tomato half over a dish or sink. You can use fingers or a spoon. Seed removal minimizes soup moisture, keeping it from getting watery.

Chop or dice tomato flesh according to recipe specifications after removing seeds. Cut them into small pieces for chunky soup or finely dice them for smoother soup.

Consider that different tomato kinds can produce varying flavors and textures in soup. Roma tomatoes thicken soups due to their low moisture content. However, beefsteak tomatoes are juicier and can bring fresh tomato taste to soup.

Fresh tomatoes add vivid, garden-fresh flavor to soup. Blanching, peeling, and deseeding tomatoes ensures a smooth, velvety soup without tomato skin bitterness or seed juice. These techniques will help you maximize your fresh tomatoes and prepare a delicious soup, whether you're creating a tomato bisque or a veggie stew.

Herbs and Greens:

Use fresh herbs to add vibrant and aromatic flavors to soups. Fresh herbs like basil, parsley, cilantro, and others can brighten even the simplest soups. How to treat them carefully:

Selection: Choose fresh, bright herbs. Look for crisp, unwilted leaves. For freshest herbs, select them from your own or buy them at a market.

Before usage, gently rinse herbs under cold water to eliminate dirt and residue. Pat them dry with a clean kitchen or paper towel. Wet herbs dilute flavors and make chopping difficult.

Chopping: Chopping fresh herbs for soup at the last minute is crucial. Cut herbs release their essential oils and tastes since they are sensitive and volatile. Use a sharp knife to finely chop herbs to distribute them evenly in the soup. Blunt knives bruise and darken plants.

Timing: Add fresh herbs to soup right before serving or after cooking. This keeps their flavors fresh and prevents wilting and scent loss. Remaining soup heat will softly infuse it with herb flavor.

For aesthetics, garnish! Sprigs of fresh herb on top of soup before serving provide color and communicate to diners that they're in for a tasty treat.

Leafy greens like spinach and kale add nutrients and texture to soups. How to handle them to make your dish shine:

Greens should be crisp and colorful. Avoid withered or yellow leaves. Fresh produce is often available at farmer's markets and gardens.

Cleaning: Gently rinse the greens under cold water to remove dirt and sand. Separate spinach leaves and soak them in a big dish of water. Lift them out of the water after swishing to dislodge particles. Keep doing this until the water clears.

Drying: Spin the leaves or gently pat them dry with a clean kitchen towel or paper towels after washing. Too much water dilutes soup flavor and consistency.

Chopping: Chop leafy greens roughly or rip them into bite-sized pieces, depending on your soup. Chop heartier greens like kale without the rough stems for a tenderer texture.

For fragile leafy greens like spinach, add them near the end of cooking. This preserves their color and nutritional content without overcooking and mushyness. Just a few minutes of boiling should do.

Fresh herbs and lush greens may make a simple soup a masterpiece. These ingredients bring color, nutrients, and flavor to your soup. When making a lovely bowl, appreciate your herbs and greens. Taste buds will appreciate you!

Roasting Vegetables:

Roasting and sautéing can improve the flavor of vegetables and soup. Each strategy will be examined in detail:

Roasting vegetables in an oven at 400°F (200°C) or higher is common. This process caramelizes the vegetables' inherent sugars, increasing their sweetness and providing taste that other cooking methods can't match. How to roast soup vegetables:

Prep: Choose your vegetables. Roasting vegetables like bell peppers, tomatoes, carrots, onions, garlic, squash, and root vegetables like potatoes and parsnips is common. To cook evenly, wash, peel (if preferred), and chop into uniform pieces.

Season the chopped veggies with olive oil, salt, pepper, and any herbs or spices. This step is essential for flavor enhancement. Cumin, paprika, chili powder, and fresh herbs like rosemary or thyme work well.

Roasting: Spread seasoned vegetables equally on a parchment- or silicone-lined baking sheet. For perfect roasting, leave space between pieces. Pre-heat the oven and roast the vegetables until soft and caramelized. Variety and size of veggies affect roasting time, however it usually takes 20-40 minutes.

Checking: Stir vegetables occasionally throughout roasting to achieve even cooking and caramelization. The finished product should be fork-tender and golden-brown.

After roasting, let the vegetables cool before adding them to soup. This lets their flavors settle and deepen and prevents overcooking in boiling liquid.

Sautéing Vegetables:

Sautéing includes frying veggies in a hot pan with little oil or butter. Searing the vegetables immediately locks in their inherent tastes and balances textures. How to sauté soup vegetables

Selection and Preparation: Like roasting, wash, peel, and chop veggies into your desired form. Onions, bell peppers, mushrooms, zucchini, and leafy greens sauté nicely.

Heat a skillet or sauté pan over medium-high heat with a little oil or butter. Olive, canola, or clarified butter work well. Heat the oil until it shimmers but doesn't smoke.

Sauté: Add prepped vegetables to heated pan. To achieve consistent cooking, stir often with a spatula or wooden spoon. The vegetables caramelize beautifully while retaining their flavor due to high heat and regular movement.

Season sautéed vegetables with salt, pepper, and herbs or spices. At this point, garlic and shallots offer taste complexity.

Most vegetables cook in 5-10 minutes when sautéed. They should be soft but crisp.

After sautéing, let the vegetables cool before adding them to the soup. A quick chilling period lets their flavors blend and develop.

Roasting and sautéing enhance vegetable flavors, making them more potent. They provide your homemade soup a deep, nuanced flavor that will impress you and your guests. Whether you roast or sauté, these methods will elevate your soup.

When making clear or delicate soups, blanching and shocking can improve the quality and look of vegetables. The vegetables are blanched briefly before shocking to quickly chill them. This method is explained in greater detail:

Blanching: A large pot of water should be brought to a boil. Use a saucepan large enough to fit your vegetables without overcrowding to cook them evenly.

Chefs may add a good spoonful of salt to boiling water. Although optional, this can improve vegetable flavor.

While the water heats, prepare the vegetables. Trim broccoli florets into bite-sized pieces and cut asparagus similarly. Remove any dirt and contaminants.

Blanch Vegetables: Use a slotted spoon or wire mesh basket to delicately lower the vegetables into the boiling water. Avoid hot water splashes. The vegetables should be in boiling water.

Blanching veggies takes 1–3 minutes, depending on type and size. Aim to soften and brighten them without entirely cooking them. Taste, texture, and color can be lost by overcooking.

Bring an Ice Bath: While the vegetables blanch, fill a big bowl with ice water. For the stunning step.

Shocking: After blanching, use a slotted spoon or tongs to take the vegetables from the boiling water and place them in the ice water bath. The vegetables stop boiling quickly in frozen water.

Let the vegetables chill in the ice water for about the same time as they were blanched, or until they are fully cool. This keeps them bright and sharp.

Why Shock and Blanche?

Blanching and shocking have many benefits:

Boiling and shocking "lock in" veggies' bright colors. This is especially important for clear or delicate soups that highlight the components.

Blanching and shocking vegetables make consommé and light vegetable broths seem clean and appetizing.

Blanched and shocked veggies can elevate clear or delicate soups. This method makes your vegetables look beautiful, flavorful, and tender, making every spoonful a treat.

Don't Rush

Don't Rush: Soup vegetables require love. Enjoy the process and be patient. The vegetables you choose, clean, and chop will affect the flavor and look of your soup."

Slowing down in the kitchen can make your cooking outstanding. This idea is especially true while cooking soup vegetables. Why patience and detail matter:

1. Choose Vegetables: Take your time in the vegetable department at the grocery store or farmers' market. Examine each vegetable carefully. Look for bright colors, firm textures, and freshness. Seasonal vegetables are cheaper and tastier.

2. Cleaning and Washing: Prior to cutting and dicing, give veggies proper care. Washing them thoroughly removes dirt and pollutants and preserves the flavor of your soup. Rinse leafy greens several times because dirt collects between their leaves.

3. Knife Skills: Vegetable cutting techniques greatly affect soup texture and cooking time. Rushing this step may result in overcooked or mushy veggies. Take your time to master knife skills for accurate, consistent cuts. This makes your soup look better and cooks evenly.

4. Flavor Development: Preparing veggies involves enhancing their flavor, not just adding them to the pot. Slowly sautéing onions until golden brown gives soup a rich, sweet flavor. Roasting veggies before adding them to the pot boosts their sweetness. These methods take time, but the soup has complex flavors.

5. Nutritional content: In addition to taste and appearance, slow vegetable cooking can maintain nutritional content. Monitoring and timing are essential to avoid vitamin and mineral loss from overcooking.

6. Mindful Cooking: Meditation can be found in preparing soup vegetables. You connect with your meal as you interact with each item. This attention improves your cooking and lets you appreciate fresh, entire ingredients.

7. Presentation: Remember, we consume with our sight and taste buds. A simple soup can become gourmet with proper vegetable handling and presentation. A soup or bowl with skillfully placed veggies can look as delicious as it tastes.

In essence, slowing down to prepare soup vegetables is culinary devotion. This honors the ingredients, respects the process, and creates a soup that nourishes the body and soul. Slow down, enjoy your time in the kitchen, and let your love for soup shine through in every spoonful.

Preparing Spices

Adding the right spices can take your soup to the next level and make it one of the best you've ever had. Spices are the unsung heroes of the kitchen, transforming even the most boring soup into something spectacular. Let's learn the ins and outs of spicing up your soup together.

The first step is to collect all of the spices you'll need. Garlic, onion powder, paprika, cumin, coriander, and thyme are just few of the basic spices that can be used in a variety of soups. Salt and pepper are the foundation of seasoning, so be sure to use them.

You could use pre-ground spices, but if you have the time and the entire spices on hand, grinding them yourself is a much better option. The fresh perfume of ground spices fills a home with pleasure as they are used for the first time. Like an aromatic spa treatment for your stomach. You can use a traditional mortar and pestle or a modern electric spice grinder to grind your spices. Throw in some whole spices and give it a whirl. The end outcome will be a flavorful spice blend. In addition, the flavor of spices is enhanced when they are freshly ground rather than when they are pre-processed.

Now, here's a pro tip for when you're ready to add these spices to your soup: sprinkle them in gradually. Instead, sprinkle them in slowly while the onions and garlic are cooking. This allows the spices to open up and impart their aromas and flavors into the soup.

And while we're on the subject of sautéing, it's important to note that warming spices in a tiny quantity of oil or butter can help bring out their full flavor. Don't be bashful about putting your spices in the pot first. A little toasting is fine, but blackened spices might make your soup taste harsh, so keep an eye on them.

Last but not least, remember that moderation is key when it comes to the spice level. More can be added at any time, but removed items cannot be replaced. So, use a small amount at first, give the soup a taste, and then add more if necessary. Trying out new things and getting creative while cooking is half the joy.

Making soup is a lot like composing a symphony of tastes and aromas from various spices. It's an adventure in scent, flavor, and trying new things. Don't be afraid to experiment a little, have faith in your tastebuds, and let the magic of well-prepared spices bring out the best in your soup.

Preparing Pulses

Cooking pulses is a crucial step in many soup recipes since they increase the soup's substance, flavor, and nutritional value. Beans, lentils, and chickpeas are examples of pulses that are not only tasty but also very nutritious. To help you make the most of these delicious soup fixings, here is a little tutorial:

First, spread the pulses out on a clean kitchen surface and remove any bad beans, discoloration, and small stones or debris that may have made their way into the bag. Then, rinse the beans thoroughly. After you've sorted your pulses, place them in a fine-mesh sieve or colander and rinse them under cold running water to get rid of any remaining debris.

Soaking: (Not required) Soaking pulses before cooking them can cut down on cooking time and make them easier to digest, but it's not always essential. Soaking can be done in two ways:

Soak Pulses in Cold Water for 12 Hours To soak your sorted and rinsed pulses overnight, place them in a big basin and cover them with cold water. Soak them over the night, or at least 8 hours.

In a pinch, you can employ the rapid soak technique. Put your dried beans and other pulses in a pot and cover them with water. Remove from heat after two to three minutes, cover, and allow stand for one to two hours.

If the pulses were soaked, drain them and give them a good rinsing in cold water. Some of the natural chemicals that can induce gastrointestinal distress are flushed out in this process.

To prepare pulses for cooking, place them in a large saucepan or pressure cooker, whether they have been soaked or not. Put in enough liquid so that they are submerged by at least a few inches. To boost the flavor, you can add aromatics such as onions, garlic, or bay leaves. Get it boiling, then turn it down to a low simmer.

Cooking Time: About 15 minutes in a simmering pot. Pulses require varying amounts of time to cook, depending on the variety and whether or not they were soaked. It could take as little as 20 minutes or as long as two hours. Taste a few now and then to see whether they're done. They need to be pliable without being mushy.

Seasoning: Add salt to taste at the end of cooking time. Pulses might become tough if salted too early in the cooking process.

Cooked pulses should be drained in a colander or sieve before proceeding with the recipe. If your soup recipe calls for part of the cooking liquid, you can set it aside.

Now that your pulses are ready, you may include them into whatever soup you like. Soups, whether they be minestrone, chili, or lentil, are poised to become a newfound gastronomic delight and healthy staple. Feel free to try out a wide range of pulses and spices in your soups.

Basic Broths and Stocks

1. Vegetable Broth

Ingredients:

- Two chopped carrots
- two chopped celery stalks
- one diced onion
- three minced garlic cloves
- one sliced leek
- one chopped parsnip
- and one bay leaf
- 1 fresh sprig of thyme
- 1 fresh rosemary sprig
- Peppercorns, one teaspoon
- Approximately 8 cups of water To taste with salt

Nutritional Values (per serving,)

Calories: 45
Protein: 1g
Carbohydrates: 11g
Fiber: 3g
Sugars: 4g
Fat: 0g

Number of Servings: 4

Instructions:

- Put a tablespoon of olive oil in a big pot and heat it over medium.
- Throw in some garlic, onion, and leek. After about 5 minutes in the pan, they should be fragrant and transparent.
- Cut up some carrots, celery, parsnips, a bay leaf, some thyme, some rosemary, and some peppercorns and throw them all into the saucepan. Allow the vegetables to simmer for another 5 minutes while you stir everything together.

- Make sure the vegetables are submerged by adding 8 cups of water. Get this to a rolling boil.

- When the broth comes to a boil, lower the heat to a simmer and let it there for at least 45 minutes and up to an hour. By doing so, the broth's flavors will develop and deepen.

- After a few minutes of simmering, take the saucepan off the heat and set it somewhere cold. Strain the broth through a fine mesh strainer or cheesecloth into a new container, throwing away the solids.

- Add salt to taste to your veggie broth. You can start with a teaspoon and add more or less to your taste.

- Soups, stews, and other dishes that call for broth will benefit greatly from your own veggie broth. It keeps well in the fridge for up to 5 days, or it can be frozen for extended storage.

2. Chicken Broth

Ingredients:

- One complete chicken, weighing around 3-4 pounds, preferably of organic origin.
- The recipe calls for two carrots that have been chopped,
- two celery stalks that have also been chopped
- one onion that has been quartered
- four cloves of garlic that have been smashed
- and two bay leaves.
- The quantity of peppercorns ranges from 10 to 12.
- One teaspoon of salt, or an amount adjusted according to personal preference.
- Approximately 10 cups of water were used.

Nutritional Values (per serving,)

Calories: 50
Protein: 6g
Carbohydrates: 3g
Fat: 2g

Number of Servings: 8

Instructions:

- Cook the Chicken: First, give your chicken a good rinse under cold water and take off the giblets. Toss the chicken into a stockpot of adequate size.

- Carrots, celery, onion, smashed garlic, bay leaves, peppercorns, and salt should be added to the chicken in the saucepan before cooking.

- Pour enough water to completely cover the chicken and veggies (approximately 10 cups).

- Place the pot over medium heat, and bring the contents to a boil. After the water has come to a boil, lower the heat to a simmer, cover the pot, and let it for at least two hours.

- Remove any scum or froth that rises to the surface while simmering. A clearer broth is the result of doing this.

- Once the chicken has cooked through and is sliding off the bone, it is time to drain the soup into a big bowl or another saucepan. Throw away the solids.

- Chicken broth should be refrigerated after it has cooled, preferably in sealed containers. It keeps well in the fridge for up to a week, and in the freezer for up to a few months.

3. Beef Broth

Ingredients

- 1 pound (450g) of beef skeletons
- Two sliced carrots, two chopped celery stalks
- one quartered onion, three smashed garlic cloves
- and a single bay leaf
- Four or five unmilled peppercorns
- A pinch of salt
- Add enough water to barely cover the ingredients.

Nutritional Values (per serving,)

Calories: Approx. 100 kcal
Protein: 8g
Carbohydrates: 5g
Fat: 5g

Number of Servings: 4

Instructions:

- In a preheated oven set to 400 degrees Fahrenheit (200 degrees Celsius), roast the beef bones for 30 minutes. The broth's flavor will be greatly enhanced by this process.
- Throw the roasted beef bones, vegetables (carrots, celery, onion), herbs (bay leaf, peppercorns), and garlic (smashed cloves) into a large stockpot.
- Add enough water to completely submerge the contents of the bowl. Don't put too much in the pot.
- Put the pot on the stove and bring the water to a boil over medium heat. As soon as it begins to boil, turn the heat down to low and simmer, partially covered, for at least 4 hours. If you want a deeper taste, boil it for longer.
- If any contaminants rise to the surface while simmering, remove them.
- The broth can be strained through cheesecloth or a fine-mesh sieve into a different pot or a large dish once it has simmered. Throw away the solids.

- Add salt to taste before serving the broth. Keep in mind that salt may always be added after the fact, but it's much harder to fix a broth that's too salty.

- Soups, stews, and other dishes that benefit from a savory broth foundation can now be started with your homemade beef broth. Any leftover broth can be refrigerated for up to 5 days or frozen for later use.

4. Bone Broth

Ingredients

- Bones from 2 pounds of meat or chicken
- 2 sliced carrots
- 1 chopped onion 2 chopped celery stalks
- Four crushed garlic cloves
- There are 2 bay leaves in this.
- Apple cider vinegar, one tablespoon
- To taste, with salt and pepper
- The ingredients should be submerged in water.

Nutritional Values (per serving,)

Calories: 50
Protein: 8g
Carbohydrates: 4g
Fat: 1g
Fiber: 1g

Number of Servings: 8

Instructions:

- To begin, give the bones a quick roast in the oven at 400 degrees Fahrenheit (200 degrees Celsius) for about 30 minutes, or until they reach a desired color. The flavor of your bone broth will be enhanced by this.

- Place the roasted bones, carrots, celery, onion, garlic, bay leaves, and apple cider vinegar in a large stockpot or slow cooker.

- The ingredients should be completely submerged in the water you pour over them. For the finest outcomes, it's crucial that everything be completely submerged.

- Prepare to taste with salt and pepper. Don't go crazy with the salt just now; you can always add more afterwards.

- Bring the ingredients to a moderate simmer in a covered pot or slow cooker. This may take some time on the cooktop, but the even heat is well worth the wait.

- Simmer the bones for at least 8 hours and up to 24 if possible. The longer it cooks, the more flavor is developed.

- Remove any froth or contaminants that rise to the surface as the mixture simmers.

- Once the soup has simmered, pour it into a big basin or container using a fine-mesh screen or cheesecloth. Throw away the solids.

- Refrigerate the broth after it has cooled. The fat will rise to the top as it cools and solidify, making it simple to skim off before serving.

- Bone broth can be used as a base for soups and stews, or simply reheated for a warming, nourishing drink.

5. Seafood Stock

Ingredients

- One pound of seafood scraps, including shrimp and crab shells, fish heads, and other odds and ends.
- 1 medium onion, chopped
- Roughly slice 2 carrots
- 2 chopped celery stalks
- Two crushed garlic cloves
- The equivalent of one bay leaf
- Fresh thyme, one sprig
- 1 fresh parsley leaf
- 1 teaspoon of ground black pepper
- 8 glasses of water

Nutritional Values (per serving,)

Calories: 25
Fat: 0g
Carbohydrates: 5g
Protein: 1g
Fiber: 1g
Sugar: 2g

Number of Servings: 8

Instructions:

- First, get rid of any grit or other debris from your seafood by rinsing it under cold running water.
- A few tablespoons of oil can be heated over medium heat in a large stockpot. Put in the vegetables that have been chopped. They need around 5 to 7 minutes in the pan to soften and turn golden brown when sautéed.
- Crush some garlic, thyme, parsley, and black peppercorns, and throw them all into the pot. Sauté for a further 2 minutes to let the flavors combine.

- Put in the pot the washed fish scraps and shells. For about 5 minutes, while tossing occasionally, sauté them with the vegetables.

- Put the 8 cups of water in the pot and heat it until it boils. Turn the heat down low after it begins to boil, then leave it to simmer uncovered for 45 minutes to an hour. The stock will become infused with the flavors of the fish.

- Using a fine-mesh strainer or cheesecloth, drain the stock into a new, larger pot or container once it has simmered. Squeeze the solids to get as much juice out of them as you can.

- You can store the seafood stock in the fridge for up to three days or freeze it in smaller pieces for later use.

How to Store and Use Leftover Broth Pureeing

Storing Leftover Broth:

Before you can start pureeing, you'll need to store your leftover broth properly to ensure its freshness. Here's what you should do:

1. Allow it to Cool: Let the leftover broth cool to room temperature. Hot liquids can cause condensation in storage containers, which can lead to spoilage.

2. Portion into Containers: Divide the broth into manageable portions. Small, airtight containers or ice cube trays work well for this purpose.

3. Label and Date: Always label your containers with the type of broth and the date it was made. This makes it easy to identify and use them later.

4. Refrigeration: If you plan to use the broth within a few days, store it in the refrigerator. It can stay fresh for up to 4-5 days under proper refrigeration.

5. Freezing: For longer storage, freeze the broth. Use freezer-safe containers or resealable bags, leaving some room for expansion. Properly stored, broth can last for several months in the freezer.

Using Pureed Broth:

Now that you have stored your leftover broth, it's time to make it even more versatile through pureeing:

1. Thaw (if frozen): If your broth was frozen, thaw it in the refrigerator overnight or gently warm it in a saucepan when you're ready to use it.

2. Blend or Puree: Transfer the broth to a blender, food processor, or use an immersion blender directly in the pot. Puree until smooth. You can also add other ingredients like cooked vegetables, herbs, or spices to enhance the flavor and texture.

3. Incorporate into Dishes:

Sauces: Use pureed broth as a base for gravies, pan sauces, or creamy pasta sauces. It adds depth and flavor while helping to thicken the sauce naturally.

Soups: Boost the flavor of your soups by adding pureed broth. Whether you're making a creamy soup or a classic chicken noodle, pureed broth can take your recipe to the next level.

Risottos and Grains: Substitute some or all of the cooking liquid with pureed broth when making risotto, rice, quinoa, or couscous. It infuses the grains with a savory taste.

Marinades: Pureed broth can serve as a flavorful marinade for meats, poultry, or tofu. It imparts a savory and aromatic essence to your dishes.

Stews and Braises: Use pureed broth as a braising liquid for meats and vegetables, creating tender and flavorful dishes.

4. Adjust Seasoning: Remember to taste and adjust the seasoning of your dishes after adding pureed broth, as its flavor can vary based on the original broth's ingredients and seasonings.

5. Store Unused Portion: If you don't use all the pureed broth at once, you can store the remainder in the refrigerator for a few days or freeze it in small portions for later use.

Pureeing leftover broth is a culinary hack that enhances your cooking repertoire while minimizing food waste. It's a sustainable and delicious way to make the most out of your homemade broths and elevate your dishes to new heights. So, next time you have leftover broth, don't discard it; puree it and savor the difference!

CHAPTER 2

Quick and Easy Soups

When it comes to mealtime, there is frequently a need for something that is not only tasty but also quick to prepare. This is because mealtimes are often rushed. To come to the rescue in these situations are speedy and straightforward soups. These reassuring bowls of goodness provide a warm and hearty supper in a short amount of time, making them a go-to alternative for times when you are pressed for time during the week or when hunger strikes without warning.

Quick and Easy Soups' Magic

Prepare effortlessly without spending hours in the kitchen. Simple soups save the most time. A substantial supper may be prepared in 30 minutes or less with a few ingredients and simple instructions.

1. Quick soups are adaptable at their best. By changing ingredients, you can generate many flavors and sensations. Traditional chicken noodles, creamy tomato bisque, and robust veggie minestrone are also available.

2. Healthy and Wholesome: Fast soups are not necessarily unhealthy. Fill them with vegetables, lean proteins, and whole grains for a healthy, guilt-free supper. You control the ingredients, so you can accommodate certain diets.

3. Comfort in Every Spoonful: Few things compare to a hot cup of soup. During cold evenings, a simple soup provides comfort and warmth, whether you're sick or just hungry.

Perfecting Quick and Easy Soups

Making delicious fast soup is as easy as 1-2-3:

1. Choose Your Base: Start with a tasty chicken, vegetable, or seafood stock. In a hurry, you can use store-bought broth or bouillon cubes.

2. Add goodies: Mix in vegetables, proteins, and aromatics. For added taste, add spices and herbs. You can use leftover roasted chicken, canned beans, frozen veggies, or fresh herbs.

3. Simmer and Serve: Let flavors blend in the saucepan. Simmering times vary by ingredient, but fast soups usually take 15-20 minutes. Serve hot with fresh herbs, olive oil, or grated cheese when everything is cooked to perfection.

Quick and Easy Soup Ideas

Tender chicken, vegetables, and slurp-worthy noodles make classic chicken noodle soup the ideal comfort dish.

Tomato Basil Soup: Creamy tomato soup with fresh basil, excellent with grilled cheese.

Minestrone: This Italian staple is wholesome and filling with a rainbow of veggies, beans, and pasta.

Thai Coconut Soup: Get a taste of Southeast Asia with coconut milk, lemongrass, and spices.

A zesty Mexican-inspired soup with tortilla strips, avocado, and lime for a flavor explosion.

Soups are your go-to for rapid cooking. A steaming bowl of comfort may be made quickly and pleasant with a little creativity and good ingredients. Next time you're short on time or need a warm lunch, try soups. Taste buds feel embraced by them.

Best Classic Soups

There are few things better than a steaming bowl of traditional soup to warm the soul. These classic dishes have been enjoyed by people of all backgrounds and cultures for many years. Classic soups are an integral part of American cuisine, from grandma's kitchen to the finest restaurants. In this article, we'll sample some of the world's finest traditional soups.

1. Classic Chicken Noodle Soup

Ingredients

- Two cups of chicken stock
- 1 cup of cooked chicken breast diced
- Carrots, diced (1/2 cup)
- Ingredients: 1 cup egg noodles, 1/2 cup sliced celery, 1/2 cup diced onion
- 1 teaspoon of minced garlic
- One Bay Leaf
- To taste, with salt and pepper.

Nutritional Values (per serving)

Calories: 180
Protein: 15g
Carbohydrates: 15g
Fat: 6g
Fiber: 2g

Number of Servings: 4

Instructions:

- Garlic, onion, carrots, and celery should be sautéed together in a big saucepan until they soften.
- Put in some chicken cubes, a bay leaf, and some chicken broth. Simmer for 20 minutes after bringing to a boil.
- When the egg noodles are done cooking, add them.

- Prepare to taste with salt and pepper.
- Comfort yourself with a nice cup of chicken noodle soup.

2. Hearty Tomato Basil Soup

Ingredients

- 4 large ripe tomatoes, diced
- 1 medium onion, finely chopped
- 3 cloves garlic, minced
- 2 cups vegetable broth
- 1/4 cup fresh basil leaves, finely chopped, plus extra for garnish
- 2 tablespoons olive oil
- Salt and black pepper, to taste
- 1/2 cup heavy cream (optional)
- Grated Parmesan cheese for serving (optional)

Nutritional Values (per serving)

Calories: 220
Protein: 4g
Carbohydrates: 20g
Fat: 16g
Fiber: 4g

Number of Servings: 2

Instructions:

- In a large pot, heat the olive oil over medium heat. Add the chopped onion and minced garlic, sautéing for about 5 minutes until they become translucent and fragrant.

- Add the diced tomatoes to the pot, stirring well to combine with the onion and garlic. Cook for an additional 5 minutes to allow the tomatoes to soften.

- Pour in the vegetable broth, bringing the mixture to a gentle boil. Reduce the heat and let it simmer for about 15-20 minutes, allowing the flavors to meld together.

- Remove the pot from the heat and let it cool slightly. Then, using a hand blender or a regular blender, carefully puree the soup until it reaches a smooth consistency. If using a regular blender, you may need to do this in batches.

- Return the pureed soup to the pot and reheat over low-medium heat. Stir in the chopped basil, salt, and black pepper, adjusting the seasonings to taste.

- If using, add the heavy cream and stir well, cooking for an additional 2-3 minutes until the soup is heated through.

- Serve hot, garnished with extra basil leaves and a sprinkle of grated Parmesan cheese, if desired.

- Enjoy this comforting Tomato Basil Soup with a side of crusty bread for a hearty and delightful meal!

15-Minute Soups

Greetings, individuals with a passion for soup! It is understood that individuals may encounter periods of busyness in their lives, necessitating the consumption of expeditious and soothing sustenance. This is where our compilation of the most optimal 15-minute soups proves to be advantageous. These recipes exhibit not only a high level of palatability but also a remarkable ease of preparation, particularly when time is limited. Let us proceed directly to the topic at hand.

1. Creamy Tomato Basil Soup

Ingredients.

- Two cups of tomato soup base.
- The quantity of heavy cream required is 1/2 cup.
- A quantity of 1/4 cup of freshly harvested basil leaves.
- Season with salt and pepper according to personal preference.

Nutritional Values (per serving)

Calories: 250
Protein: 5g
Carbohydrates: 15g
Fat: 20g
Fiber: 3g

Number of Servings: 4

Instructions

- The tomato soup base should be heated in a saucepan over medium heat until it reaches a simmering point.
- Decrease the temperature to a lower setting, incorporate the heavy cream into the mixture, and continue stirring until thoroughly blended.
- Carefully separate the freshly picked basil leaves into smaller fragments and include them into the soup. Gently agitate the mixture.
- Add salt and pepper according to personal preference.

- Continue to simmer the mixture for an additional duration of 5 minutes, ensuring to stir intermittently.

- Consume while the food is at an elevated temperature and derive pleasure from the experience.

- It is possible to enhance the taste of your soup by adding a small amount of grated Parmesan cheese or croutons as a garnish.

2. Chicken Noodle Soup

Ingredients.

- A quantity of chicken broth equivalent to four cups.
- Two cups of shredded, cooked chicken breast.
- One unit of measurement equivalent to 1 cup of egg noodles.
- One carrot that has been cut into small, uniform pieces.
- One stalk of celery, finely chopped.
- One-half of an onion, which has been finely diced.
- Season with salt and pepper according to personal preference.
- Fresh parsley can be used as a garnish.

Nutritional Values (per serving)

Calories: 300
Protein: 25g
Carbohydrates: 20g
Fat: 10g
Fiber: 2g

Number of Servings: 6

Instructions

- In a sizable container, heat the chicken broth until it reaches its boiling point.
- Incorporate the finely chopped carrot, celery, and onion. Allow the mixture to simmer for a duration of five minutes, or until such time that the vegetables have reached a state of tenderness.

- Incorporate the shredded chicken and egg noodles into the mixture.
- Continue cooking for an additional duration of 5 minutes, or until the noodles have reached a state of tenderness.
- Add salt and pepper according to personal preference.
- Transfer the contents into individual bowls, adorn with freshly chopped parsley, and present the dish while it is still steaming.

5 Ingredients-Soups

1. Leek and Potato Chowder

Ingredients:

- Peel and dice 4 big potatoes.
- Two leeks, sliced very thinly
- 4 cups broth, either chicken or vegetable
- 12 cups of the thick stuff
- To taste, with salt and pepper

Nutritional Values (per serving)

Calories: 250
Protein: 5g
Carbohydrates: 35g
Fat: 10g

Number of Servings: 4

Instructions:

- Mix potatoes, leeks, and broth together in a big pot. Obtain a rolling boil.
- Simmer over a low heat for 15 minutes, or until potatoes are cooked through.
- The soup can be pureed with an immersion blender.
- Salt and pepper the dish, then add the heavy cream and stir.
- Continue to simmer, stirring periodically, for another 5 minutes.
- Enjoy the ease of preparation while still hot.

2. Two-Bean Soup with Spinach

Ingredients:

- 1 small onion, diced finely
- 2-inches of minced garlic
- White beans, one 15-ounce can (after draining and rinsing)

- Veggie stock, about 4 cups
- Leafy greens, to the tune of 4 cups
- Garnish with Parmesan cheese (if desired)

Nutritional Values (per serving)

Calories: 180
Protein: 8g
Carbohydrates: 30g
Fat: 2g

Number of Servings: 4

Instructions:

- The onions and garlic should be cooked until transparent in a big pot.
- White beans and veggie broth should be added. Get it to a low boil.
- Cook until wilted, then stir in the fresh spinach leaves.
- You can partially puree the soup with an immersion blender while still leaving some beans intact.
- Garnish with Parmesan cheese and serve immediately if desired.

CHAPTER 3

Vegetarian Soups

1. Soup with Lentils and Vegetables

Ingredients:

- Green lentils, dry, one cup
- Vegetable stock, four cups
- 1 diced onion
- 2 diced carrots and 2 diced celery stalks
- 2 minced garlic cloves
- Diced tomatoes, one 14-ounce can
- 1/2 tsp coriander
- Paprika, 1/2 teaspoon
- To taste, with salt and pepper
- The use of freshly chopped parsley as a garnish

Nutritional Values (per serving)

Calories 180
Protein: 2g
Carbohydrates: 45g
Fiber: 6g
Fat: 1g
Servings: 42g

Number of Servings: 6

Instructions

- Prepare a bowl of cold water to soak the lentils in.
- Dice the onion, carrot, and celery and sauté them with a little olive oil in a big pot until they soften.

- Cook for one more minute, or until the garlic, cumin, and paprika have released their aroma.

- Add the rinsed lentils, diced tomatoes (with their juice), and vegetable broth to the pot.

- The lentils should be cooked after 20 to 25 minutes of simmering the sauce at a low boil.

- Prepare to taste with salt and pepper.

- Serve immediately by ladling into dishes and topping with parsley.

2. Soup with Butternut Squash

Ingredients:

- 1 cubed and peeled butternut squash
- 2 apples that have been peeled, cored, and diced 1 chopped onion
- Vegetable stock, four cups
- Cinnamon, one teaspoon's worth
- 12 tsp. of ground nutmeg
- To taste, with salt and pepper
- Garnish with sour cream or Greek yogurt if you like.

Nutritional Values (per serving)

Calories: 180
Protein: 2g
Carbohydrates: 45g
Fiber: 6g
Fat: 1g

Number of Servings: 4

Instructions

- To make translucent onion, sauté the chopped onion in a large pot.
- Cook the butternut squash and apples for a few minutes until they soften to your liking.

- Put the veggie broth in the pot and heat it until it boils.

- Simmer for 20–25 minutes, or until the apples and squash are soft.

- The soup can be pureed with an immersion blender. You can also use a conventional blender, but do so only after the mixture has cooled slightly.

- Add some salt, pepper, cinnamon, and nutmeg for flavor.

- Depending on your taste, top with sour cream or Greek yogurt and serve hot.

Vegan Soups

1. Creamy Butternut Squash Soup

Ingredients

- Cubed, seeded, and skinned butternut squash (one medium-sized)
- 2 minced garlic cloves 1 chopped onion
- Vegetable stock, four cups
- One 13.5-ounce can of coconut milk
- Olive oil, enough for 2 tablespoons
- Season with salt and pepper to taste, then top with fresh sage leaves if desired.

Nutritional Values (per serving)

Calories: 220
Protein: 3g
Carbohydrates: 28g
Fat: 12g
Fiber: 4g

Number Servings: 6

Instructions:

- Olive oil should be heated over medium heat in a big saucepan.
- Throw in some garlic and onions, chopped. To get the onions to become translucent, sauté them for four minutes.
- Prepare a vegetable broth and add butternut squash cubes. Squash should be soft after 15 minutes of simmering at a low heat after being brought to a boil.
- The soup can be easily made smooth using an immersion blender. You can also put it in a blender, but you should wait until it cools off a bit before doing so.
- If you feel the need to do so, return the soup to the pot and add the coconut milk. Prepare to taste with salt and pepper.
- To finish heating, simmer for another 5 minutes.
- Garnish with fresh sage leaves, if desired, and serve hot

2. Chickpea and Spinach Soup

Ingredients

- Two drained and rinsed cans of chickpeas (15 ounces each)
- one sliced onion
- Two finely chopped garlic cloves
- Vegetable stock, four cups
- Two cups of baby spinach leaves
- Olive oil, enough for 2 tablespoons
- A teaspoon of cumin seed powder
- To taste, with salt and pepper
- Garnish with lemon slices (if desired).

Nutritional Values (per serving):

Calories: 260
Protein: 12g
Carbohydrates: 36g
Fat: 8g
Fiber: 10g

Servings: 4

Instructions:

- Olive oil should be warmed over moderate heat in a big saucepan.
- Toss in some minced garlic and onions. When sautéed, onions should be tender and fragrant, after around 3 minutes. Add the ground cumin and continue cooking for one more minute.
- Put in some chickpeas and some vegetable stock. Bring to a boil, then lower heat and simmer for 10 minutes.
- Half of the soup should be pureed using an immersion blender; the chunky chickpeas should be left in. After about 2 minutes, stir in fresh spinach and heat until wilted.
- Use salt and pepper to taste.
- Hot, with a squeeze of lemon juice, if desired.

Protein-Packed Options

1. Grilled Chicken Salad

Ingredients:

- 2 chicken breasts, skinned and boneless
- Greens (lettuce, spinach, arugula, etc.) to taste, around 4 cups
- 1 cup of halved cherry tomatoes
- Half a red onion, sliced thinly 1 cucumber, sliced
- Feta cheese crumbles, about a quarter cup
- Dressing with balsamic vinaigrette

Nutritional Values (per serving)

Calories: 350
Protein: 30g
Carbohydrates: 10g
Fat: 20g
Fiber: 4g

Number of Servings: 2

Instructions:

- Turn on the grill to medium heat. Sprinkle some salt, pepper, and olive oil over the chicken breasts.

- For an internal temperature of 165 degrees Fahrenheit (75 degrees Celsius) and clear juices, grill the chicken for about 6 to 8 minutes per side.

- Mixed greens, cherry tomatoes, cucumber, and red onion should be tossed together in a big bowl while the chicken grills. Let the chicken rest for a few minutes after it's cooked before slicing it thinly.

- Add the sliced chicken and feta cheese crumbles to your salad. Balsamic vinaigrette dressing can be drizzled on top. Combine all of the ingredients and serve at once.

- Make a salad your own by adding avocado, nuts, or your preferred vegetables as toppings.

2. Quinoa and Black Bean Bowl

Ingredients:

- One cup of rinsed and drained quinoa
- Two ounces of water
- Black beans, one 15-ounce can (after draining and rinsing)
- maize kernels, one cup (fresh or frozen)
- One chopped red bell pepper
- One-half cup of finely chopped fresh cilantro
- To One lime's juice.
- Ground cumin, one teaspoon
- Taste with salt and pepper.
- A garnish of avocado slices would be nice.

Nutritional Values (per serving)

- Calories: 400
- Protein: 15g
- Carbohydrates: 70g
- Fat: 5g
- Fiber: 12g

Number of Servings: 4

Instructions:

- In a saucepan, bring the water to a boil, then add the quinoa. Reduce heat, cover, and simmer for about 15 minutes or until quinoa is cooked and water is absorbed.
- In a large bowl, combine the cooked quinoa, black beans, corn, diced red bell pepper, and chopped cilantro.
- In a small bowl, whisk together lime juice, ground cumin, salt, and pepper. Pour the dressing over the quinoa mixture and toss to combine.
- Serve in bowls, garnished with avocado slices if desired.

Chapter 4

Meat Lovers' Soups

1. Beef and Barley Soup

Ingredients:

- 1 pound of beef stew meat, cubed
- 1/2 cup of pearl barley
- 1 onion, chopped
- 2 carrots, sliced
- 2 celery stalks, chopped
- 4 cups of beef broth
- 2 cloves of garlic, minced
- 1 teaspoon of dried thyme
- Salt and pepper to taste

Nutritional Values (per serving)

Calories: 350
Protein: 25g
Carbohydrates: 30g
Fat: 12g
Fiber: 6g

Number of Servings: 6

Instructions:

- Oil should be heated in a big pot over moderate heat. Brown the meat cubes on all sides after adding them to the pan.
- Take the meat out of the equation. Chopped onions, carrots, and celery should be added to the same pot. Saute until softened, about 5 minutes.
- Prepare with garlic mince and thyme. For maximum aroma, stir for a full minute.

- Put the steak back in the pot and stir in the pearl barley.

- Put the beef broth in and heat it until it boils.

- Simmer for 45 minutes to an hour, covered, until the beef is soft and the barley is cooked.

- Taste and adjust salt and pepper.

- Eat while it's hot and enjoy the succulent pork flavor!

2. Sausage and Potato Chowder

Ingredients:

- A pound of cut sausage links
- Cubed potatoes equaling 4 cups
- One medium-sized onion, chopped 2 garlic cloves, minced
- Chicken stock, four cups
- One cup of full-fat cream
- Shredded cheddar cheese, one cup
- Taste with salt and pepper.
- Green onions, chopped, for garnish

Nutritional Values (per serving)

Calories: 400
Protein: 15g
Carbohydrates: 30g
Fat: 25g
Fiber: 4g

Number of Servings: 4

Instructions:

- Sausage slices should be fried in a big pot. Take out and put to one side.

- Onion and garlic should be sautéed in the same pot until aromatic.

- Mix with some potato dice and chicken stock. Simmer for 15 minutes, or until potatoes are cooked, after bringing to a boil.

- Add the cooked sausage together with the cream and cheese. Cook over low heat until the cheese melts.
- Taste and adjust salt and pepper.
- Serve the sausage and cream soup in dishes and top with chopped green onions.

CHAPTER 5

Classic Seafood Soups

1. New England Clam Chowder

Ingredients:

- Minced clams, two 10-ounce cans
- Chopped bacon equivalent to 4 slices
- Diced potatoes, one cup
- Onions, diced, half a cup
- One-half cup of chopped celery
- Four ounces of cream
- A quarter of a cup of full-fat milk
- Butter, two tablespoons.
- Taste with salt and pepper.
- Garnish with chopped fresh parsley.

Nutritional Values (per serving)

Calories: 350
Protein: 10g
Carbohydrates: 25g
Fat: 24g
Fiber: 2g

Number of Servings: 4

Instructions:

- To make crispy bacon, sauté chopped bacon in a big pot over medium heat. Take the bacon out of the pot but keep the grease in there.
- Dice some onions and celery and throw them in the saucepan. To make them transparent, sauté them in bacon fat.

- Mix in the diced potatoes and clams (along with whatever juice they came in). Set the timer for five minutes.

- Bring the mixture to a simmer after adding the whole milk and heavy cream. The potatoes should be cooked in about 10 to 15 minutes in the simmering liquid.

- Mix in the melted butter and bacon. Taste and adjust salt and pepper.

- Serve immediately while still steaming hot; spoon into dishes and top with minced parsley.!

2. Shrimp and Corn Chowder

Ingredients:

- Medium-sized shrimp (about 1 pound): skin and devein.
- Two cups of corn kernels, frozen
- Diced red bell peppers, one cup
- 1 cup of finely sliced onion 2 minced garlic cloves
- Chicken stock, four cups
- Half and Half, one cup
- Butter, 2 Tablespoons
- Old Bay seasoning, 1 tablespoon
- Taste with salt and pepper.
- Garnish with some fresh chives.

Nutritional Values (per serving)

Calories: 300
Protein: 25g
Carbohydrates: 30g
Fat: 10g
Fiber: 3g

Number of Servings: 6

Instructions:

- Butter should be melted in a big pot over low to medium heat. Throw in some garlic, onion, and pepper flakes. Saute until soft, about 5 minutes.

- Add the frozen corn and Old Bay seasoning and stir to combine. Prepare in 5 minutes.

- Add the chicken broth and reduce the heat to low.

- Toss in the shrimp and cook for three to four minutes, or until the flesh is opaque and a pink color.

- Combine the half-and-half with the seasonings of your choice and stir.

- Serve immediately by ladling into dishes and topping with chives..

Exotic Seafood Soups

1. Thai Coconut Shrimp Soup

Ingredients:

- Peeled and deveined 1 pound of big shrimp
- Coconut milk, one 14-ounce can
- Soup stock, either chicken or vegetable, to taste, 2 cups
- Chopped and crushed two lemongrass stalks
- Two or three slices of galangal or ginger
- two or three kaffir lime leaves, torn
- Depending on how spicy you like it, thinly slice two red Thai chilies.
- fish sauce, 1 tablespoon
- 1 tbsp. of dark sugar
- Extract from 1 lime
- Cilantro leaves, fresh for the table

Nutritional Values (per serving)

Calories: 300
Protein: 20g
Carbohydrates: 10g
Fat: 20g
Fiber: 2g

Number of Servings: 4

Instructions:

- Coconut milk and broth should be combined in a big saucepan and heated over medium.
- Mix in the kaffir lime leaves, red Thai chilies, galangal, or ginger. Simmer it for around five to seven minutes to bring out the flavors.
- Mix the fish sauce and brown sugar together until smooth.
- After about 4 minutes, add the shrimp and boil them until they turn pink.

- Lime juice is the finishing touch, so squeeze some in and mix.

- Take out the kaffir lime leaves, lemongrass, and galangal or ginger.

- Garnish with chopped fresh cilantro and serve hot. Experience the authentic Thai cuisine.

- Please note that the heat can be controlled by the amount of red Thai chilies used.

2. Spanish Seafood Paella Soup

Ingredients:

- Arborio rice, one cup
- Clean and debearded mussels, half a pound
- 1/2 pound of sanitized clams
- Large shrimp, about a half pound's worth, peeled and deveined
- Onion, diced, 1/2 cup
- Red bell pepper, diced, 1/2 cup
- 2 minced garlic cloves
- Saffron threads, enough for one teaspoon
- Smoked paprika, one teaspoon
- Four cups of fish or chicken broth Two tablespoons of olive oil
- Taste with salt and pepper.
- Prepare a garnish with some fresh parsley.
- Sliced lemons for garnishing

Nutritional Values (per serving):

Calories: 400
Protein: 20g
Carbohydrates: 45g
Fat: 15g
Fiber: 3g

Number of Servings: 4

Instructions:

- Olive oil should be heated over medium heat in a big skillet. Put in the minced garlic, onion, and red bell pepper. Heat in a pan till tender.

- Mix in the smoked paprika and saffron threads.

- After 1–2 minutes, add the Arborio rice and toss to coat the grains in the spice mixture.

- Add the broth and reduce heat to low. Cook for 15–20 minutes covered, or until liquid is absorbed and rice is soft.

- Put the shrimp, clams, and mussels in the pot. The seafood is done cooking after the shells of the mussels and clams have opened and the pot has been covered for an additional 7 minutes of cooking time.

- Taste and adjust salt and pepper.

- Garnish with parsley and lemon wedges and serve immediately. Get a taste of Spain right here!

CHAPTER 6

Fall Soups

1. Butternut Squash and Apple Soup

Ingredients:

- One cubed and peeled butternut squash
- Two apples, prepped by peeling, coring, and dicing
- 1 onion cut very small
- Two minced garlic cloves
- Vegetable stock, four cups
- One-half teaspoon of cinnamon powder
- Season with salt and pepper to taste. Sauté in olive oil

Nutritional Values (per serving)

Calories: 150
Protein: 2g
Carbohydrates: 35g
Fat: 1g
Fiber: 5g

Number of Servings: 6

Instructions:

- In a large saucepan, heat a few tablespoons of olive oil over moderate heat.
- Sauté the garlic and onion until the onion is transparent and aromatic.
- Mix in the chopped apples and butternut squash. Set the timer for five minutes.
- Combine the ground cinnamon, salt, and pepper, and add to the dish.
- Put in the veggie broth and heat till boiling.
- Squash should be cooked for 20–25 minutes at a low simmer, covered.

- The soup can be pureed using an immersion blender.
- Try it out, and if it needs more salt, add it.
- Sprinkle with freshly cracked pepper and olive oil, and serve hot.

2. Creamy Wild Mushroom Soup

Ingredients:

- One pound of sliced wild mushrooms (such shiitakes, creminis, and oysters)
- one onion, cut very small
- 2 minced garlic cloves
- Vegetable stock, four cups
- One cup of full-fat cream
- Butter, 2 Tablespoons
- Garnish with fresh thyme leaves.
- Taste with salt and pepper.

Nutritional Values (per serving)

Calories: 250
Protein: 5g
Carbohydrates: 10g
Fat: 20g
Fiber: 2g

Number of Servings: 4

Instructions:

- Melt the butter in a large saucepan over medium heat. Add the chopped onion and garlic, sauté until softened.
- Toss in the sliced mushrooms and cook until they release their moisture and start to brown. Pour in the vegetable broth and bring to a simmer. Let it cook for 10-15 minutes.
- Use an immersion blender to puree the soup until creamy. Stir in the heavy cream and season with salt and pepper. Simmer for an additional 5 minutes.
- Ladle into bowls, garnish with fresh thyme leaves, and serve piping hot.

CHAPTER 7

Winter Soups

1. Creamy Butternut Squash Soup

Ingredients

- Medium-sized butternut squash, prepped by peeling, seeding, and cubing.
- Ingredients: 1 medium onion; 2 minced garlic cloves
- 2-cups of vegetable stock
- 1/2-gallon light cream
- Add 2 tbsp. of olive oil
- Adjust seasonings to taste with salt and pepper.
- Decorate with some fresh thyme leaves.

Nutritional Values (per serving)

Calories: 300
Protein: 3g
Carbohydrates: 30g
Fat: 20g
Fiber: 5g

Number of Servings: 6

Instructions:

- Olive oil should be heated in a big pot over moderate heat. Sauté the garlic and onion until the onion is transparent and aromatic.
- Cook the butternut squash in a pot with some veggie broth. Simmer for about 15 minutes, or until the squash is soft, after bringing to a boil.
- You can smooth up the soup using either a standard blender or an immersion blender.

- Heavy cream should be added to the soup when it is returned to the pot. Keep at a low boil for another 5 minutes.
- Taste and adjust salt and pepper.
- Garnish with fresh thyme leaves and serve hot.

Note: For an extra kick of flavor, consider adding a pinch of nutmeg or a dash of cayenne pepper.

2. Beef and Barley Soup

Ingredients:

- 1 cup of barley 1 pound of cubed beef stew meat
- 2 diced carrots
- 2 diced celery stalks
- 1 chopped onion
- 4 cups beef stock
- 1 bay leaf
- Salt and pepper to taste
- Prepare a garnish with some fresh parsley.

Nutritional Values (per serving)

Calories: 350
Protein: 25g
Carbohydrates: 40g
Fat: 10g
Fiber: 7g

Number of Servings: 4

Instructions:

- The beef stew meat should be browned in a large pot over medium heat until it is evenly browned on all sides.
- Put in the pot the chopped onion, celery, and carrots. To soften the vegetables, sauté them for a few minutes.

- Add the barley, bay leaf, and beef broth.
- To cook the barley and beef, bring to a boil, then reduce heat and simmer for about 30 minutes.
- Put in as much salt and pepper as you like.
- Take out the bay leaf, divide into bowls, and sprinkle with parsley.
- favorites like mushrooms and peas that thrive in the colder months.
- Indulge in one of these hearty soups this winter to stay toasty and full. These dishes will satisfy your cravings for anything from smooth butternut squash to substantial meat and barley. Enjoy!

CHAPTER 8

Spring Soups

1. Asparagus and Pea Soup

Ingredients

- Trimmed and cut fresh asparagus, equivalent to 1 bunch
- Peas, 2 cups (either fresh or frozen)
- one onion, cut very small
- 2 minced garlic cloves
- Vegetable stock, four cups
- 1/4 cup of full-fat milk
- 1.5 ounces of olive oil
- Taste with salt and pepper.
- To garnish, use fresh mint leaves (optional)

Nutritional Values (per serving)

Calories: 200
Protein: 5g
Carbohydrates: 18g
Fat: 12g
Fiber: 6g

Number of Servings: 4

Instructions:

- Olive oil should be heated in a big pot over moderate heat. After about 2 minutes, add the onion and garlic and cook until the onion is transparent and the garlic has released its aroma.
- Chop some asparagus and some peas and throw them in the pot. Keep cooking for another three to four minutes.

- Put in the veggie broth and heat till boiling. Simmer over a low heat for 10–12 minutes, or until the asparagus is cooked through.

- The soup can be pureed with an immersion blender or a conventional blender (in batches).

- Heavy cream should be added to the soup when it is returned to the pot. Gently heat, but do not let to boil

- Taste and adjust salt and pepper as needed.

- Garnish with fresh mint leaves and serve hot.

- Note: Crusty bread or grated Parmesan cheese go wonderfully with this soup.

2. Spinach and Lemon Orzo Soup

Ingredients:

- Pasta orzo, one cup
- Four cups of fresh spinach leaves six cups of vegetable broth
- The zest and juice of one lemon
- Half onions are cut very small
- Two minced garlic cloves
- Half an ounce of olive oil
- Taste with salt and pepper.
- Optional fresh dill garnish.

Nutritional Values (per serving)

Calories: 250
Protein: 5g
Carbohydrates: 35g
Fat: 9g
Fiber: 3g

Number of Servings: 6

Instructions:

- Olive oil should be heated in a big pot over moderate heat. After about three or four minutes of cooking time, add the onion and garlic.

- Orzo pasta should be added and roasted for a couple of minutes, stirring continuously.
- The veggie broth should be added and brought to a boil. Simmer the orzo for 8-10 minutes, covered, until it reaches the desired consistency.
- Toss the fresh spinach in and stir it until it wilts.
- To taste, add salt and pepper and then add the lemon juice and zest.
- Garnish with fresh dill if you like and serve hot.

CHAPTER 9

Cold Soups for Hot Days

1. Chilled Cucumber Gazpacho

Ingredients:

- Four peeled and diced cucumbers
- Two minced garlic cloves One diced red bell pepper Half chopped red onion
- Three cups of tomato juice
- 1/4 cup of balsamic vinegar
- olive oil, extra virgin, 1/4 cup
- a single teaspoon of salt
- Black pepper, half a teaspoon
- Decorate with some fresh dill.

Nutritional Values (per serving)

Calories: 120
Protein: 2g
Carbohydrates: 12g
Fat: 8g
Fiber: 2g

Number of Servings: 6

Instructions:

- In a blender or food processor, combine the cucumbers, red bell pepper, red onion, and garlic. Blend until smooth.
- Add the tomato juice, red wine vinegar, olive oil, salt, and black pepper. Blend again until everything is well combined.
- Chill the gazpacho in the refrigerator for at least 2 hours before serving.
- When ready to serve, garnish with fresh dill and a drizzle of olive oil.

- Enjoy the cool and refreshing flavors!

Pro Tip: For an extra kick, add a dash of hot sauce or a pinch of cayenne pepper if you like a bit of heat in your gazpacho.

2. Chilled Watermelon Soup

Ingredients:

- Six ounces of chopped ripe watermelon
- One-half cup of mint leaves
- Lime Juice, Fresh, 1/4 Cup
- Honey, to taste, 1 tbsp
- A quarter of a teaspoon of black pepper 1/2 teaspoon of salt
- Adding Greek yogurt as a topping

Nutritional Values (per serving)

Calories: 80
Protein: 1g
Carbohydrates: 20g
Fat: 0g
Fiber: 1g

Number of Servings: 4

Instructions:

- In a blender, combine the watermelon cubes, fresh mint leaves, lime juice, honey, salt, and black pepper.
- Blend until you achieve a smooth, liquid consistency.
- Chill the watermelon soup in the refrigerator for at least 1 hour before serving.
- Serve in bowls or glasses, garnished with a dollop of Greek yogurt and a sprig of fresh mint.
- Sip and savor the sweet and tangy flavors of summer!

CHAPTER 10

World Soups

1. Tom Yum Goong - Thai Spicy Shrimp Soup

Ingredients:

- Shrimp, peeled and deveined, weighing about seven ounces (200 grams), with four cups of chicken or shrimp broth
- Pieces from Two lemongrass stalks
- Three to four slices of galangal or ginger
- There should be three or four kaffir lime leaves.
- 200 grams (7 ounces) of sliced mushrooms 2 or 3 red bird's-eye chilies, bruised
- Two to three tablespoons of fish sauce
- 2–4 tbsp. of fresh lime juice
- Cilantro leaves, fresh for the table

Nutritional Values (per serving)

Calories: 150
Protein: 15g
Carbohydrates: 10g
Fat: 5g
Fiber: 2g

Number of Servings: 4

Instructions

- Chicken or shrimp broth should be brought to a boil in a pot.
- Put in some bird's eye chilies, kaffir lime leaves, lemongrass, galangal, or ginger.
- Keep simmering for 5-10 minutes to develop the flavor.
- Cook the shrimp and mushrooms until the shrimp become pink, about 3 minutes.

- Prepare to taste by adding fish sauce and lime juice.
- Garnish with chopped fresh cilantro and serve hot.

2. Gazpacho - Spanish Cold Tomato Soup

Ingredients:

- One cucumber, peeled and diced; 6 ripe tomatoes, diced
- 1 chopped red bell pepper
- Finely chopped red onion from 1 small onion
- two minced garlic cloves
- Juice from 3 tomatoes, 3 cups
- 14 cups of balsamic vinegar
- Toss with 1/4 cup olive oil
- Season with salt and pepper to taste. Garnish with fresh basil or parsley.

Nutritional Values (per serving)

Calories: 120
Protein: 2g
Carbohydrates: 12g
Fat: 7g
Fiber: 2g

Number of Servings: 6

Instructions:

- In a blender, combine diced tomatoes, cucumber, red bell pepper, red onion, and garlic.
- Blend until smooth.
- Transfer the mixture to a large bowl, and stir in tomato juice, red wine vinegar, and olive oil.
- Season with salt and pepper to taste.
- Chill in the refrigerator for at least 2 hours.
- Serve cold, garnished with fresh basil or parsley.

CHAPTER 11

Healing and Medicinal Soups

1. Turmeric Ginger Immunity Booster Soup

Ingredients:

- Vegetable stock, four cups
- Grated fresh turmeric root equal to 1 tablespoon
- Minced fresh ginger, one tablespoon
- Two minced garlic cloves
- A single cup of sautéed spinach, carrots, and bell peppers.
- One cup of ready-to-eat quinoa
- Taste with salt and pepper.
- Adding fresh lemon juice gives it a tangy kick.

Nutritional Values (per serving)

Calories: 180
Protein: 6g
Carbohydrates: 35g
Fat: 2g
Fiber: 5g

Number of Servings: 4

Instructions:

- The vegetable broth should be simmered over medium heat in a large pot.
- Put in the ginger, garlic, and turmeric that have been finely chopped. For the next five minutes, simmer the soup with the herbs.
- Add the cooked quinoa and the mixed vegetables. Cook at a low simmer until the greens are fork-tender.
- Add some salt, pepper, and freshly squeezed lemon juice, to taste.

- Enjoy the restorative powers of this bright, immunity-boosting soup by serving it hot.

2. Elderberry and Honey Soothing Soup

Ingredients:

- Approximately 3 cups of water
- Elderberries, about a half cup's worth
- Raw honey, 2 tablespoons
- Stick of cinnamon, one
- Ground cloves, just a little
- Add some lemon zest for a zesty kick.

Nutritional Values (per serving)

Calories: 70
Protein: 0.5g
Carbohydrates: 19g
Fat: 0g
Fiber: 4g

Number of Servings: 6

Instructions:

- In a pot, combine the water, dried elderberries, cinnamon stick, and ground cloves.
- Bring to a boil, then reduce the heat and let it simmer for 20 minutes.
- Remove from heat and strain the liquid into a bowl.
- Stir in the raw honey and lemon zest.
- Allow the soup to cool slightly before serving.
- Sip on this soothing elixir to ease cold and flu symptoms or simply enjoy it as a comforting beverage.

BONUS CHAPTER

Best Breads for Your Bowl

1. Garlic Parmesan Rolls

Ingredients:

- Four pillow dinner rolls
- Two finely chopped garlic cloves
- Butter, melted; 2 teaspoons
- Parmesan cheese, grated, 2 tablespoons
- Optional fresh parsley garnish.

Nutritional Values (per serving - 1 roll):

Calories: 180
Protein: 4g
Carbohydrates: 20g
Fat: 9g
Fiber: 1g

Number of Servings: 4

Instructions:

- Have a 375F (190C) oven ready.
- Split each roll in half across the width.
- Combine the garlic powder and melted butter in a small bowl.
- Spread the garlic butter on the rolls' sliced surfaces.
- Grate some Parmesan cheese and sprinkle it on top of the buns.
- Bake the rolls for 5 to 7 minutes, or until the bread is golden and the cheese is melted and bubbling.
- Fresh parsley, if preferred, can be used as a garnish.

- Astound your guests by serving them these savory buns alongside your soup.

2. Whole Grain Sourdough

Ingredients:

- 1 loaf of whole grain sourdough bread

Nutritional Values (per serving - 2 slices):

Calories: 200
Protein: 8g
Carbohydrates: 40g
Fat: 2g
Fiber: 6g

Number Servings: 6 (12 slices total)

Instructions:

- Cut the sourdough bread made with whole grains into slices that are one and a half millimeters thick.

- Toasted to a golden brown and just to the point of being crisp, the slices should be.

- These thick slices are excellent for dipping into your favorite chunky soups or for constructing an amazing grilled cheese sandwich to go with your tomato soup. Both options are wonderful for a cold winter night.

Salads that Complement

1. Classic Caesar Salad

Ingredients:

- A head of romaine lettuce, ripped into smaller pieces and set aside.
- Croutons up to half a cup
- a quarter of a cup's worth of grated Parmesan cheese
- A quarter cup of the dressing used for Caesar salad
- To taste, salt and black pepper will be used.
- It's up to you. Protein options include grilled chicken or shrimp.

Nutritional Values (per serving, without protein)

Calories: 250
Protein: 5g
Carbohydrates: 15g
Fat: 18g
Fiber: 3g

Number of Servings: 4

Instructions:

- In a large salad bowl, add the torn romaine lettuce.
- Sprinkle the croutons and grated Parmesan cheese over the lettuce.
- Drizzle the Caesar salad dressing on top.
- Season with a pinch of salt and a generous grind of black pepper.
- Toss everything together until the dressing evenly coats the salad.
- If desired, add grilled chicken or shrimp for a complete meal.

2. Mediterranean Quinoa Salad

Ingredients:

- One cup of quinoa, raw or cooked

- One diced cucumber
- One cup of halved cherry tomatoes 1/2 cup of pitted, sliced Kalamata olives
- a quarter of a cup of chopped red onion; a quarter of a cup of crumbled feta
- Extra-virgin olive oil, two tablespoons.
- Fresh lemon juice, a teaspoon
- Dried oregano, one teaspoon
- To taste, with salt and pepper
- Garnish with some fresh parsley.

Nutritional Values (per serving):

Calories: 300
Protein: 7g
Carbohydrates: 30g
Fat: 18g
Fiber: 5g

Number of Servings: 4

Instructions:

- Combine the quinoa that has been cooked, the chopped cucumber, the cherry tomatoes, the Kalamata olives, the red onion, and the crumbled feta cheese in a large mixing bowl.
- Olive oil, fresh lemon juice, dried oregano, salt, and pepper should be mixed together in a small basin using a whisk.
- After pouring the dressing over the salad, toss it to evenly coat the ingredients.
- Add some freshly chopped parsley as a garnish.
- You can have it as a side dish, or you can have it as a nutritious, light supper.

When Wine is Fine

1. Grandma's Secret Sangria

Ingredients:

- One bottle of a full-bodied red wine
- a quarter of a cup of brandy
- One-fourth of a cup of orange liqueur
- 1 orange, peeled and segmented 1 lemon, peeled and segmented 1 lime, peeled and segmented 2 tablespoons of sugar (adjust to taste)
- Ice fragments

Nutritional Values (per serving)

Calories: 180
Alcohol: 14g
Sugar: 5g

Number Servings: 6

Instructions:

- Mix all of the ingredients together in a big pitcher
- Then, place it in the refrigerator for at least a few hours, but preferably overnight, to allow the flavors to meld.
- Pour over ice, then raise a glass to the good times!

2. Mediterranean Magic White Wine Spritzer

Ingredients:

- One bottle of dry white wine
- One glass of sparkling fluid
- A freshly squeezed lemon
- A few sprigs of mint.
- A few ice cubes

Nutritional Values (per serving)

Calories: 90
Alcohol: 7g
Sugar: 1g

Number of Servings: 4

Instructions:

- A taste of what the Mediterranean has to offer:
- Combine the sparkling water, lemon juice, and white wine in a mixing glass.
- After you've poured it over ice and decorated it with some fresh mint leaves,
- you may enjoy a taste of summer in a glass.

3. Berry Bliss Wine Smoothie

Ingredients:

- 1 glass of crimson wine
- half a cup of frozen berries mixed together
- half a banana
- Half a cup of yogurt, either Greek or another type of your choosing.
- One teaspoon and a tablespoon of honey
- A handful of frozen water

Nutritional Values (per serving)

Calories: 200
Alcohol: 12g
Fiber: 3g

Number of Servings: 2

Instructions

- Blend the red wine
- frozen berries, banana, yogurt, honey, and ice until smooth.
- Sip your way into a world of berry bliss with this delightful wine smoothie.

4. Zesty Lemon-Garlic Shrimp Scampi

Ingredients:

- 1 pound of large shrimp, peeled and deveined, with the shells removed
- 4 garlic cloves, chopped into small pieces
- One lemon's worth of zest and juice
- A quarter of a cup of dry white wine.
- a couple of teaspoons' worth of butter
- 2 teaspoons of extra-virgin olive oil
- To taste, salt and pepper is available.
- Parsley leaves picked fresh for the garnish.
- Pasta, cooked, of whichever variety you choose

Nutritional Values (per serving, excluding pasta):

Calories: 250
Protein: 20g
Carbohydrates: 3g
Fat: 17g

Number of Servings: 4

Instructions:

- Heat the olive oil and butter in a skillet.
- Add minced garlic and sauté until fragrant.
- Add shrimp, lemon zest, and juice.
- Cook until shrimp turn pink, then stir in the white wine.
- Season with salt and pepper, garnish with fresh parsley, and serve over cooked pasta.

5. Roasted Red Pepper & Tomato Wine Soup

Ingredients:

- Two roasted and skinned red bell peppers
- Diced tomatoes, one 14-ounce can
- Dry red wine, one cup
- One small onion, diced
- Two finely chopped garlic cloves
- One milliliter of olive oil
- Paprika, smoked, 1/2 teaspoon
- Basil leaves for garnish Salt and pepper to taste

Nutritional Values (per serving)

Calories: 120
Alcohol: 5g

Number of Servings: 4

Instructions:

- Olive oil should be heated till shimmering before adding the onion and garlic.
- Paprika, roasted peppers, diced tomatoes, and wine should all be added at this point.
- After simmering for 15 minutes, the mixture should then be blended until smooth.
- After seasoning with salt and pepper and garnishing with basil, you may then relish the flavorful soup that has been infused with wine

TIPS AND TRICKS

Perfect Soups Every Time

1.Start with Quality Ingredients: Creating a truly excellent soup requires a solid foundation, which can be laid by beginning with high-quality components. It's really similar like painting a masterpiece; in order to make something beautiful, you need to use the best possible paints and brushes. In the world of cooking, the ingredients you select can be compared to the materials you use in your artwork, and here is why this comparison is so important:

- Seasonal Fruits and Vegetables: The choice of veggies you make for your soup will determine its flavor and texture. Choose produce that is as fresh and vivid as possible whenever you can. Not only does freshness contribute to a superior flavor, but it also contributes to the dish's overall nutritional content. When you utilize vegetables that are at their peak maturity and in season, you'll notice that the finished dish retains the inherent richness and depth of the ingredients. Therefore, visit your neighborhood market in search of the most appetizing-appearing vegetables, or give some thought to cultivating some of your own for an unparalleled farm-to-table experience.

- Rich and Flavorful Broths: The broth is the foundation of many different types of soups, and the quality of the broth determines how successful your creation will be. Even though making your own broth from scratch is amazing if you have the time to do so, there are still excellent options for purchasing broth in stores. Try to find broths that contain a low amount of sodium and have few additives. Your soup will have more layers of taste and scent if the broth you use is of high quality, which will also make it more satisfying. It's similar to adding a high-quality wine to a sauce; doing so takes the dish to an entirely new level.

- Proteins That Have Been Properly Aged: If your soup contains proteins like chicken, beef, or shellfish, the quality of those proteins plays a significant role in the overall flavor. Before adding your meats to the soup, put some attention into seasoning them and preparing them. To bring out their full potential, try marinating, searing, or roasting them. Your soup will be infused with a delectable flavor and a pleasant mouthfeel if you include high-quality proteins in it, whether those proteins are juicy bits of shrimp or soft chunks of chicken.

In its most basic form, the notion of beginning with high-quality ingredients isn't just about enhancing the flavor of your soup; rather, it's about developing a culinary adventure that appeals to all of your senses. When you give priority to using fresh, high-quality ingredients, your soup will transform into a piece of art that is a demonstration of the care and attention that you have placed in each individual component. If you give your ingredients the courtesy of treating them with reverence, they will repay you with a symphony of flavors in each and every bite.

2. Sweat the Aromatics: Before you add your stock or liquid, take a few minutes to sweat your aromatics, such as onions, garlic, and celery, in some oil or butter. This step is done before you add your stock or liquid. At this stage, the focus is on drawing out the latent tastes that are contained within these components. When you give these aromatic vegetables a low and slow cooking, it's almost like you're preparing the canvas for a gourmet masterpiece. The heat makes the onions more pliable, brings out the sweet and earthy aroma of the garlic, and smoothes out the astringency of the celery. What is the result? A base that is aromatic and full of taste, and it acts as the foundation of your soup.

Imagine that these scents are the first notes of a symphony, and that they are responsible for establishing the tone for the entire piece. They produce an atmosphere of harmony that serves as a base over which you can construct layers of flavor and intricacy. Therefore, afford them the attention that they require, and permit them to take center stage.

3. Build Your Flavors in Layers: Imagine the flavors in your soup to be playing a symphony together. In the same way that a wonderful piece of music is created when a variety of instruments and melodies come together to create it, layering flavors in your soup enables each component to offer its own distinctive essence.

Start with a sturdy base, which will often consist of your sweated aromatics, and then gradually add other components. This may consist of vegetables, meats, or spices, among other things. It is possible for the tastes to combine and become more harmonious if you add the ingredients one at a time and then let them boil together.

Imagine that you are adding new instruments to the symphony; each one has a specific purpose, and when they all work together, the resulting piece is one that is complex and full of nuances. This is what you should expect. Your soup will have more dimension, more going on for it, and an experience in the mouth that will stick with you.

4. Broth Made at Home vs Broth Bought in a Store: A culinary treasure, homemade broth is produced with care from scratch after being simmered for hours to extract every last bit of flavor from the bones, vegetables, and herbs used in the preparation. Making your own broth is really gratifying and may take your soup to a whole new level if you have the time and the motivation to do so.

However, life may be chaotic, and when it is, it is helpful to have options that can be purchased in a store. You shouldn't be afraid to use them, but you should use some discretion. Choose options with less sodium so you can exert more influence over the level of salty in the finished product. Also, keep in mind that even if you utilize broth that you bought from the store, you can still impart your own distinctive flavors into it by according to the layering and seasoning advice given previously

5. Season your soup with care: Patience is essential when it comes to seasoning your soup properly. Take your time with this very important step. Taste your soup as it develops throughout the course of the simmering process. Add the seasonings one at a time, waiting some time between each addition to allow the flavors to combine. Think about seasoning as though you were putting the finishing touches on a masterpiece. Beginning with a little brushstroke and tasting as you go, gradually increasing the amount until you achieve the appropriate level of flavor depth. It is important to keep in mind that you can always add more salt, pepper, or herbs, but once something is added, it cannot be removed. Take your time, listen to what your taste tells you, and you'll end up with a cup of soup that's just right.

6. Fresh Herbs at the Finish: The addition of fresh herbs such as basil, parsley, and cilantro at the end of a dish is analogous to adding the final touches to a beautiful picture. They lend your soup a vivacity, color, and an explosion of taste that is bursting with freshness. However, if you want their vibrant colors to remain intact after cooking, you should add them in the last few minutes.

Imagine that these herbs are the last notes in a piece of music, and use that analogy. They are a pleasant finishing touch that leaves a lasting impression on your taste receptors and elevate the quality of the entire encounter. Don't be scared to try out a variety of herbs to see which ones work best with the soups you make. Experimenting is the best way to learn.

7. Creaminess without Cream: Creamy soups are certainly comforting, but you can create that luscious texture without the heaviness of cream by using a stock that has been reduced in volume. You might want to try mixing in some roasted vegetables or bean puree. This clever hack not only makes the dish creamier, but it also increases the amount of nutrients and fiber it contains.

Imagine this as a miraculous shift in the array of dishes you have available to you. Your soup will have a velvety texture and a luxurious feel without the use of cream, which is high in calories. You won't have to sacrifice flavor or texture, which is a bonus, and you'll have the satisfaction of knowing you've made a healthy selection for yourself.

8. Don't Throw Away the Leftovers: Soups have a wonderful characteristic that, in many cases, allows them to get better with age. The leftovers taste even better than the freshly produced batch because the flavors combine, causing them to become more profound and intense. Don't be afraid to put together a large batch of your preferred soup and savor it over the course of many days; it will keep well in the refrigerator.

Imagine this as the experience of drinking a vintage wine that, like you, improves with age and gets more nuanced. Because of the additional time, the flavors of the components are able to meld together, making for a very wonderful eating experience. To ensure that the quality of your soup is preserved, you should always store it correctly in the refrigerator or freezer.

9. The Texture of Your Soup Is Important: The texture of your soup is an important factor in determining how much you enjoy eating it. It is not need to worry if it is too thick; just add a little bit more liquid until it reaches the appropriate consistency. On the other hand, if the consistency of your soup is off, try simmering it for a longer period of time. The longer it simmers, the more the flavors will meld together and the richer the texture will become.

Consider how you might go about sculpting an artwork while thinking about this component. You have the ability to sculpt and refine the consistency of your soup until it reaches your desired level. Pay attention to how it feels on your palate, and don't be hesitant to make changes until you get the perfect equilibrium for the dish.

10. Experiment and Have Fun: Finally, keep in mind that making soup is more of an experience than a method that must be followed precisely. In the kitchen, you shouldn't be scared to let your imagination run wild. Experimentation and happy accidents have led to the creation of some of the most delicious soup recipes.

Imagine that you are creating your very own, one-of-a-kind symphony of flavors, with yourself serving as the conductor, and the components serving as your orchestra. Have fun with your culinary masterpieces while exploring new ingredients and putting together surprising flavor combinations. The journey, not the destination, is where the fun of cooking lies; thus, let your imagination go wild and relish the delectable results of your experimenting in the kitchen.

You'll be well on your way to constructing soups that are not only healthful but also a wonderful culinary experience if you incorporate these tips and tricks into your repertoire of soup-making techniques. I hope you enjoy making soup!

Making Soup in Batches

1. **Choose the Right Recipes:** Opt for soup recipes that freeze well and won't lose their texture or flavor when reheated. Classics like chili, tomato soup, or vegetable broth-based soups often work great.

2. **Gather Quality Ingredients:** Start with fresh, high-quality ingredients. The better the ingredients, the tastier your soup will be. If possible, buy in bulk to save money when making large batches.

3. **Invest in Quality Containers:** Get yourself some sturdy, airtight containers that are freezer-safe. Mason jars, reusable silicone bags, or plastic containers with tight-fitting lids are good options. Label them with the date so you can keep track of freshness.

4. **Prep and Chop Efficiently:** When making a large batch, prep all your vegetables and ingredients beforehand. It makes the cooking process smoother and faster.

5. **Use Large Pots:** Having a big soup pot or a slow cooker is a lifesaver when

making batches. You can cook more soup at once, saving time and energy.

6. **Portion Control:** Consider portioning your soup into single or family-sized servings before freezing. This way, you can easily grab what you need and avoid thawing more than necessary.

7. **Cool Before Freezing:** Let your soup cool down to room temperature before freezing it. This prevents condensation and ice crystals from forming inside the container.

8. **Label and Date:** Always label your containers with the name of the soup and the date you made it. This helps you keep track of what's in your freezer and when it's time to use it.

9. **Thaw Safely:** When you're ready to enjoy your batch-cooked soup, thaw it in the refrigerator or use the defrost setting on your microwave. Avoid leaving it at room temperature for extended periods.

10. **Reheat with Care:** Reheat your soup gently over low to medium heat to maintain its flavor and consistency. Stir occasionally to ensure even heating.

11. **Get Creative:** Don't be afraid to get creative with your batch-cooked soups. You can always add fresh herbs, spices, or a splash of cream when you reheat them to give them a new twist.

Freezing and Reheating

- **Freezing:** Freezing is your best friend whether you are preparing a large quantity of your favorite dish, whether it be a casserole, soup, or even chili. This is the procedure to follow:

- **Relax and take it easy:** First, allow your food to reach room temperature before continuing. This helps minimize condensation in the freezer, which is one of the contributing factors that might cause freezer burn.

- **Controlling Portions:** Separate your food into pieces that are suitable for a meal. If you are using containers that are airtight or bags that are safe for the freezer, you need to ensure that there is some room for the food to expand.

- **Freeze Flat:** If you are going to use bags, you should aim to freeze them in a flattened state. This not only makes it simpler to stack them, but it also helps preserve space in the freezer.

Heating up:

When it comes time to enjoy your frozen product, there are a few important details to keep in mind, including the following:

- To safely thaw your frozen dinner, it is best to place it in the refrigerator the night before and let it sit there. This is the safest approach since it maintains a constant temperature at which your food may be consumed safely.

- Magic of the Microwave: If time is of the essence, you can always utilize the microwave. To ensure that your meal is heated uniformly, the microwave should be set to the defrost setting or the power level should be reduced.

- Reheating on the stovetop using a simmering method is an excellent choice for re-heating soups and stews. To ensure that the food is heated evenly, use a setting between low and medium heat and stir it occasionally.

- Baking in the Oven: Casseroles and other baked meals typically reheat wonderfully when placed in the oven. Cover with foil to keep the food from drying out, and cook it at a lower temperature (about 325 degrees Fahrenheit or 160 degrees Celsius) until it is fully warmed.

It's All About Steaming: Steaming is great for making items like dumplings or veggies that have been steamed. The addition of a little amount of additional moisture, along with a steamer basket or microwave-safe container with a lid, can work miracles.

Adjusting Flavors

When it comes to altering the flavors in your food, achieving the proper balance is the most important thing to focus on. The following advice will assist you in enhancing the flavor of the food that you prepare:

- **Taste as You Go:** The best way to fine-tune the flavors of a dish is to taste it at regular intervals while it is being prepared. First, give your dish a taste before you add any seasoning, and then give it another taste after every seasoning addition.

- **Salt and Pepper:** When it comes to making adjustments to the flavor, salt and pepper are your two best friends. The flavor of a food can be improved by adding a little bit of salt, and freshly ground pepper can give it a kick without being overpowering.

- **Acidic Ingredients:** Ingredients such as lemon juice, vinegar, or even just a splash of hot sauce are great ways to brighten up your dishes and balance off the sweetness or richness of other flavors.

- **Herbs & Spices:** If you want your flavors to have more nuance and dimension, play around with different herbs and spices. At the very end of the cooking process, you may add a burst of freshness to your dish by adding fresh herbs like basil, cilantro, or parsley.

Your foods' savory qualities can be elevated with the addition of umami-rich items such as soy sauce, Worcestershire sauce, or mushrooms. Umami can also be found in meat.

Cream and Dairy If the heat level of your food is too high, you can tone it down by adding a little bit of cream or dairy. This works wonderfully for making curry and other hot sauces.

Texture: Make sure you don't overlook the importance of texture. To create a contrast between the more tender components of a dish, it is sometimes necessary to include a crunchy component, such as roasted almonds or breadcrumbs.

In foods that feature contrasting flavors, such as sweet and savory, it is important to strike a balance between those sensations rather than allowing one to completely overshadow the other. Take, for instance, the combination of a sugary glaze and salty bacon.

The more you practice, the better you will become at adjusting flavors, which is why you shouldn't be scared to experiment with different combinations. After some time has passed, you will acquire a more refined understanding of the preferences of both you and your family.

Presentation and Garnish Ideas

The following are some presentation and garnishing tips that will set your dishes apart from others:

1. 1.Sprinkles of fresh herbs, such as basil, parsley, or chives, can impart a flavorful and visually appealing punch to a dish. They merely need to be finely chopped and sprinkled over the dishes right before they are served. They bring out the best in everything from pasta to soup to salad and even grilled meats.

2. Edible Flowers: The appearance of your meal can be improved by the addition of edible flowers, such as nasturtiums or pansies. They are beautiful, vibrant, and entirely edible all at the same time. You can use them as a garnish on salads, desserts, or even savory dishes if you arrange them in a smart pattern.

3. Zest of Citrus: Adding just a little bit of citrus zest to your dishes can make them more flavorful. Zest from citrus fruits like lemon, lime, or orange gives a zippy kick. For a zestier take on baked items, seafood, or salads, grate the zest and sprinkle it on top.

4. Sauce Drizzles: Learn how to become an expert in the art of sauce dripping! You can make sophisticated patterns on your plates by using a squeeze bottle or a spoon. A simple dish can be elevated to the level of a culinary masterpiece with the addition of a balsamic reduction or oil flavored with herbs.

5. Shallots or Onions that Have Been Fried Crispy Shallots or onions that have been fried crisply not only add a delightful crunch to a dish but also a punch of flavor. They provide a new dimension when sprinkled on top of salads, soups, or dishes that are influenced by Asian cuisine.

6. Microgreens: Although they are very little and delicate, microgreens contain a lot of taste and have a lot of visual appeal. Spread some of it on some sandwiches, wraps, or even scrambled eggs to give your cuisine a new and contemporary spin on things.

7. Fruit Slices: Salads, desserts, or charcuterie boards can all benefit from the addition of thin slices of fruits like apples, pears, or strawberries, which bring a burst of color and a touch of sweetness. They are very delicious when served

alongside cheese plates.

8. Consider Using Creative Plate Shapes You should give some thought to the shape of the plates you use. You can give the impression that your food is more artistic and intriguing by serving it in square, rectangular, or another unusually shaped dish.

9. Build Your Dishes Using Layers and Stacking To Create Height and Depth, Build Your Dishes Using Layers and Stacking. Desserts, lasagnas, and even salads benefit tremendously from this technique.

10. Give Your Garnish a Purpose Always keep in mind that the flavors of your food should be complemented by the garnishes you use. If you want the experience of eating to be enhanced by the garnish you choose, you should think about the flavor and texture of the garnish.

CONCLUSION

"The Soup Cookbook" is not merely a compilation of recipes; rather, it is an exploration of the vibrant and comfortable world of soups, where flavors, traditions, and creativity come together to create meals that are reassuring, nutritious, and utterly delectable. Throughout our culinary adventure inside these pages, we've explored a fascinating assortment of soup recipes, each with its own distinct personality and appeal.

During the course of our investigation into the nature of soups, we came upon a more profound fact: namely, that soups are not only a source of nutrition for the body, but also a soother for the spirit. These recipes have illuminated the alchemy behind transforming simple, commonplace ingredients into culinary marvels that have the potential to touch our hearts and bring people together, whether at the dinner table with their families or at a party with their friends.

However, "The Soup Cookbook" is more than just a collection of recipes and cooking tips. It encourages you to make changes, come up with new ideas, and put your own personal spin on each recipe.

It is important to keep in mind, as you embark on your very own soup-making expedition, that the genuine joy of cooking does not lay just in the finished meal, but rather in the entire gourmet process. The process of creating something, the aroma that fills your kitchen, and the excitement of anticipating being able to share what you've made with others you care about are all components of the enchantment. It's about making new memories with loved ones around the dinner table and relishing the feeling of coziness and sustenance that homemade soups bring into your life.

Therefore, with your ladle in one hand and your ingredients in the other, let "The Soup Cookbook" to be your constant companion as you travel forth into the world of gastronomy. The time-honored and soul-satisfying love of soup is sure to bring you and those around you closer together, and it is my hope that this book will encourage you to create numerous bowls of comfort food and spark your creativity. I hope that your path through the culinary world is filled with many more moments that warm your heart and flavors that stick with you forever. Have fun in the kitchen!

Made in the USA
Las Vegas, NV
08 January 2024

84045948R00063